Benefits and Risks of Knowledge-Based Systems

Council for Science and Society:
Reports of Working Parties
also published by
Oxford University Press

Human Procreation: Ethical Aspects of the New Techniques
UK Military R & D
Companion Animals in Society

Benefits and Risks of Knowledge-Based Systems

Report of a Working Party
Council for Science and Society

Oxford New York Tokyo
OXFORD UNIVERSITY PRESS
1989

Oxford University Press, Walton Street, Oxford OX2 6DP

Oxford New York Toronto
Delhi Bombay Calcutta Madras Karachi
Petaling Jaya Singapore Hong Kong Tokyo
Nairobi Dar es Salaam Cape Town
Melbourne Auckland

and associated companies in
Berlin Ibadan

Oxford is a trademark of Oxford University Press

Published in the United States
by Oxford University Press, New York

British Library Cataloguing in Publication Data
Data available

Library of Congress Cataloging in Publication Data
Data available

ISBN 0-19-854743-9

Printed in Great Britain by
Biddles Ltd, Guildford & King's Lynn

About the Council

The Council for Science and Society, a registered charity, was formed in 1973 with the object of 'promoting the study of, and research into, the social effects of science and technology, and of disseminating the results thereof to the public'.

The Council's primary task is to stimulate informed public discussion in the field of the 'social responsibility of the scientist'. It seeks to identify developments in science and technology whose social consequences lie just over the horizon, where no full-scale debate has yet begun. Experience shows that intensive analysis of the present (and probable future) 'state of the art', and of the foreseeable social consequences, can suggest a range of possible responses to those who will sooner or later have to take the necessary decisions. The Council carries out this task in a number of different ways, including the organization of conferences, seminars, and colloquia.

Major studies are conducted by *ad hoc* working parties, composed — like the Council itself — of experts in the respective fields together with lawyers, philosophers, and others who can bring a wide range of skills and experience to bear on the subject. The results of these studies are published in the form of reports such as this one. It is the Council's hope that these will help others to work out the most appropriate solutions to these problems in the course of responsible public debate, conducted at leisure on the best information available, rather than by hurried, ill-informed, and ill-considered process which is apt to occur if the community does not become aware of a problem until it is too late.

The Council is grateful to the Trustees of the Esmée Fairbairn Trust and the Nuffield Foundation for financial support.

The Council welcomes all suggestions for further subjects of study.

COUNCIL FOR SCIENCE AND SOCIETY,
3/4 St Andrews Hill,
London EC4V 5BY
Tel: 01-236-6723

Council Members

Membership of the Working Party

Professor Margaret Boden: Chairwoman
Professor of Philosophy and Psychology at the School of Cognitive and Computing Sciences, University of Sussex.

Dr Mike Sharples: Rapporteur
Lecturer in Artificial Intelligence, School of Cognitive and Computing Sciences, University of Sussex.

Professor Alan Bundy
SERC Senior Fellow and Professorial Fellow, Department of Artificial Intelligence, University of Edinburgh.

Professor Mike Cooley
Director of ESPRIT Project 1217 (Human Centred Systems). Visiting Professor at UMIST. Guest Professor at the University of Bremen.

Dr John Dawson
Head of Professional, Scientific and International Affairs at the British Medical Association.

Mr David Hopson
Director of UNET Information Technology Project. Part time lecturer, Kingston College of Further Education, Staff Development Unit.

Rabbi Julia Neuberger
South London Liberal Synagogue.

Mr Barry Sherman
Writer, novelist and consultant on technology and information.

Mr Paul Sieghart [†]
Honorary Visiting Professor of Law, King's College, London. Amateur computer programmer, specialising in 'expert' systems.

Mr John Churcher: Consultant
Department of Psychology, University of Manchester.

[†]Paul Sieghart died on 12 December, 1988. He was the founder of the Council for Science and Society. He originally proposed the Knowledge-Based Systems working party and was an active member of it, contributing both to the planning and drafting of this report. He informed and animated our meetings and his death is a great loss to the Council.

Preface

This report aims to alert the public to the potential benefits, and the possible dangers, of a new form of advanced information technology: knowledge-based systems. These advice-giving computer programs are beginning to be used in an ever-widening range of institutions in our society: commercial, financial, and manufacturing companies; medical, welfare, educational, and legal services; policy-making bodies such as central and local government; and, of course, the military.

To appreciate how knowledge-based systems might help or harm us, one needs to know something about how they work — What can they do, and how? What can't they do, and why? The first chapter explains what these advisory systems are and how they produce their advice. It shows how they differ from more familiar sorts of program, being able to overcome some — but not all — of the limitations popularly ascribed to computer programs. And it outlines the way in which human knowledge must be expressed if it is to be represented in a knowledge-based system.

The second chapter surveys a varied range of specific applications in the areas of finance, the law, education, the military, and health. Many of these examples are already in use, and others will be available very soon.

This new technology attracts both dedicated champions and vociferous critics. Both points of view are outlined in the third chapter, which attempts to distinguish those tasks and uses for which knowledge-based systems might be beneficial from those where their use might be inappropriate, or even dangerous. Even where they are helpful, knowledge-based systems should be regarded as tools for our use — not as decision-makers in themselves. The more their human users understand about their potential and limitations, the less will human responsibility be undermined in using them.

The design, marketing, and use of knowledge-based systems should aim to maximise the understanding and control enjoyed by the human user. In the fourth chapter, a number of suggestions are made about how these systems can be produced responsibly and used appropriately. Extra precautions have to be taken if the program is to be used not by a human expert (such as a doctor or lawyer) but by someone knowing very little about the problem-domain concerned.

The final chapter lists the conclusions and recommendations agreed by the working party. We hope that these will be discussed and developed by readers in many walks of life, for advice-giving computer programs will soon affect everyone in our society, either directly or indirectly.

While all of the working party contributed to the preparation of this report, thanks are due above all to the rapporteur, Dr Mike Sharples, who is primarily responsible for its final form. He was helped throughout by the Council's secretarial staff, especially Keith Williams, to whom we are most grateful.

Finally, the Council thanks the Renaissance Trust, whose generosity provided funds which made this report possible.

Margaret Boden, Chairwoman of the Working Party
August 1988

Contents

1 How Knowledge-Based Systems work

1.1 Knowledge-Based Systems in practice

In May 1983 the British Government launched the Alvey Programme of advanced information technology. Costing £350 million, of which £200 million has come from public funds and £150 million from industry, its aim is to stimulate research and development in information technology. The Alvey programme has four main themes, and one of these, funded by over £30 million of Government money and equivalent investment from industry, is 'Intelligent Knowledge-Based Systems'. We shall use the abbreviation KBS rather than IKBS throughout; the term embraces the more familiar 'expert systems'.

So KBSs mean business: not yet 'big business', but rapidly growing in all the industrial nations. The aim of research in KBSs is to make computers more useful, and thus more profitable, by providing them with some of the attributes of human intelligence, such as the ability to converse in a normal human language, to reason and explain the reasoning process, to interpret and describe visual images (such as those produced by a TV camera), and to form and carry out plans.

Some KBSs are already in commercial products: for example the Caisse d'Epargne Bank in France uses one at 60 of its branches to give advice on home loan applications, and Thomas Cook has a system which plans journeys on the Australian railway network and offers guidance on suitable train schedules. Some systems, such as vision units for industrial robots, are at the stage of prototypes, or are in limited use. The four 'Alvey Demonstrator Projects' give a flavour of what is being investigated.[1]

These Demonstrator Projects provide the following examples of collaborative research leading to saleable products:

- The Design to Product Demonstrator uses KBSs to help designers at all stages of product development, from initial design through manufacture to in-service support. Information gleaned at one stage is used in subsequent ones.

- The DHSS Demonstrator will produce a series of systems for DHSS officers and Social Security claimants. One will assist DHSS local officers in processing claims and help them to interpret the mass of regulations governing Supplementary Benefits. Another will advise claimants on their likely eligibility for benefits, the procedures they should follow, and the documents they need to take in support of a claim. It will also offer help in filling in the DHSS forms. A third system will assist a DHSS policy officer in investigating the consequences in a change in benefits in terms of consistency and overlap with other benefits.

- The Speech Input Word Processor and Workstation Demonstrator will produce a machine that can turn human speech into text for a word processor.

- The Mobile Information System Demonstrator is intended to bring the benefits of Information Technology to a mobile user. Its subprojects include a Mobile Electronic Office, and a Traffic Information Collator which will take police traffic incident reports (such as notification of accidents or road blockages) and, from them, prepare messages suitable for broadcasting by local radio.

You may be surprised by the limited aims of these projects. There is no 'Home Help Robot', or 'Automated Car Driver' or 'Computer-based Lawyer' amongst them. Perhaps the most far-reaching is the Speech Input Word Processor, which could transform office practices. But perfect speech recognition is not expected. The user will be able to choose between alternative interpretations, or to repeat a word or phrase. A prototype has been produced which works far more slowly than would be acceptable, takes only speech from 'an adult male reading prepared sentences in a relatively quiet acoustic environment', and provides 'up to five best guesses about the sentence actually input to the machine'. Some of these limitations will be overcome, and the prototype turned into a more general and flexible speech recognition system, but one of the lessons from the short history of KBSs has been that such generalisation is always far more difficult and time-consuming than it first appears.

1.2 Images of computers

It is a common finding of social research that people believe computers to be more intelligent than they really are. For instance, in a survey of a college statistics class in 1981, 54 per cent of the class said a computer could beat the world chess champion.[2] The reasons for this mass delusion provide a good introduction to the social and political implications of KBSs.

Even as they were building the world's first computers, researchers were imagining the possibility of designing 'artificial intelligences'. Being academics, they saw intelligence very much in academic terms: the ability to reason, to play games like chess, and to solve problems, rather than to sing, to dance, or to see. As details of the early computers hit the headlines, so the researchers' enthusiasms were embellished and conveyed to papers eager for some respite from gloomy post-war news. In 1946 Lord Mountbatten, as President of the Institution of Radio Engineers, gave a speech, reported in *The Times* under the headline 'Electronic Brain', in which he talked about the development of:

> an electronic brain, which would perform functions analogous to those at present undertaken by the human brain. It would be done by radio valves, activating each other in the way that brain cells do...now that the memory machine and the electronic brain were upon us, it seemed that we were really facing a new revolution; not an industrial one, but a revolution of the mind, and the responsibilities facing the scientists today were formidable and serious.[3]

So the phrases 'electronic brain' and computer 'revolution of the mind' were born. In 1956 John McCarthy, a leading computer scientist, organised a 'two month ten-man study of artificial intelligence' at Dartmouth College, New Hampshire, and Artificial Intelligence (AI) became a recognised academic discipline.

1.3 Research in Artificial Intelligence

A consequence of the wilder speculations of AI practitioners (for instance that 'artificial intelligence will equal and surpass human mental abilities — if not in twenty years then surely in fifty')[4] and of portrayal of 'thinking machines' in the guise of familiar technology (such as HAL in the film *2001, A Space Odyssey*) is that the popular image of KBSs is now a blur of fact and fantasy.

Many of the researchers in AI hold the hope that general purpose thinking machines will one day be built, but present-day research goals are far more modest: to develop computational techniques (such as formalisms for reasoning or representing knowledge) with the potential for emulating intelligence, to understand the human mind by building a computer model of some aspect of intelligence (such as logical reasoning) and to design programs that match or beat human performance on a carefully defined task (such as diagnosing diseases of the blood, or converting spoken language into written text).

When it comes to automating the processes of reasoning, AI has had most success in the area of deduction, i.e. the kind of reasoning involved in proving mathematical theorems in a formal logic. Even here success has been limited to fairly straightforward theorems, due to the problems of controlling the search for a proof through the explosively large space of admissible reasoning steps. Forays have also been made into the areas of uncertain reasoning, reasoning by analogy, and a few other kinds of 'plausible reasoning', but these are only in their infancy.

In representing knowledge, AI has had some success in the representation of properties and relationships between simple physical objects. However, a large number of tricky areas remain which are only beginning to be understood, such as the representation of shape, time, liquids, and the beliefs of others. In addition, there is the sheer scale of the problem. No existing system contains more than a minute fraction of the knowledge, especially the common-sense knowledge, known to the average human.

This has limited AI systems to small, self-contained domains in which fairly crude reasoning is adequate. Early AI work concentrated on 'toy' domains, the most famous of which was the 'blocks world', a domain consisting of children's bricks in which the principal action was stacking. Until the mid-70s AI researchers were pretty depressed about this. It was felt that the initial predictions about 'thinking machines' had proved to be hyperbole, that they had only scratched the surface of the problem, and that major breakthroughs were required. This is still true.

What changed the climate of opinion from depression to enthusiasm was the realisation that many commercially important domains were 'toy' ones, in the sense suggested above. This led to the development of 'expert' systems. For the most part expert systems use automated reasoning techniques that were already well understood ten to twenty years ago, applied to a narrow domain of specialised expertise in areas such as fault diagnosis. They are not capable of the kind of common-sense reasoning about shape and time involved in, say, navigating a car in a busy street, nor of jumping to plausible conclusions about the intentions of an adversary, nor of using analogy to apply a wide range of previous experience to a new problem.

1.4 Inside expert systems

An expert system is a computer program that performs in a similar way to a human, in some limited area of expertise, and that makes use of stored representations of knowledge as a means of providing solutions and explanations. It is usually designed with the help of human experts who solve a range of test problems and describe their thought processes as they proceed. The programmer attempts to transfer the experts' knowledge and methods of reasoning into a form that can be processed by a computer program. Hayes-Roth [5] identifies seven common characteristics of expert systems:

- They can solve very difficult problems as well as or better than human experts.

- They reason heuristically, using what experts consider to be effective rules of thumb and they interact with humans in appropriate ways, including natural language.

- They manipulate and reason about symbolic descriptions.

- They can function with data that contain errors, using uncertain judgemental rules.

- They can contemplate multiple competing hypotheses simultaneously.

- They can explain why they are asking a question.

- They can justify their conclusions.

Although this list captures the important aspects of expert systems, it is a wishful rather than accurate description of present-day programs. Their 'natural language' is severely restricted (often they accept only single words or short phrases from the human user) and the explanations they provide are often unnatural, inappropriate and verbose (listing all the rules used in reaching a conclusion, rather than just those of interest to the human). Nevertheless, for certain tasks, such as diagnosis and fault-finding, expert systems can offer useful advice. They are most effective as knowledgeable assistants to a human expert (who will have a far greater range of abilities, including plain common-sense). A more appropriate term than 'expert system' is the Japanese phrase 'job assistant system'.

Here is a dialogue with a very simple expert system which gives advice on the choice of wine for a meal:

```
Welcome to the Wine Advisor.
I can help you choose the best wine for your meal.

Answer the question either with a single word from the list
offered or a single word followed by a certainty
in the range 0 to 1, eg: red certainty 0.7

Is the main dish fish, meat, or poultry?
    > meat
What type of meat: pork, veal, lamb, etc
    > veal
What type of sauce is with the main dish: tomato, creamy, spicy?
    > creamy
Is the sauce sweet or sour?
    >                       <the user does not specify a choice>
Do you prefer red or white wine?
    > red certainty 0.6
How sweet do you like your wine: dry, medium, sweet?
    > medium
How strong tasting is the meal: delicate, average, strong?
    > delicate
What kind of body would you like the wine to have: light,
    medium, full?
    > medium


I would suggest a Riesling (certainty 0.4 )
Would you like another suggestion (y/n)
    > y
I would suggest a Soave (certainty 0.4 )
Would you like another suggestion (y/n)
    > n
Consultation finished
```

Most existing expert systems have four main components:

- A database of facts, also known as a 'knowledge-base'. These represent what the system 'knows' about the problem at any one moment; for example in the Wine Advisor the knowledge-base may at some point contain the entries:

```
wine colour unknown
```

```
main dish   meat certainty 1.0
main dish contains veal certainty 1.0
```

The state of the knowledge-base will change during the consultation, as the system gains new facts. Many different formalisms for representing knowledge have been devised, from the simple assertions like those shown above, to ones which involve grouping together small 'chunks' of knowledge into larger units that are meant to emulate the organisation of concepts in the human mind.

- A set of rules that operate on the facts. A typical rule in the Wine Advisor program might be:

```
Rule colour1
IF the colour of the wine is unknown
AND the main dish is meat
AND the main dish does not contain veal
THEN
remove the fact that the colour is unknown
add the fact that the colour of the wine should be red,
giving it a certainty of 0.9
```

The rule is shown here in English; for the expert system it is written in a more formal notation, but one that is quite understandable to a human reader. For example, the last line would be typed in as:

```
add([wine colour red certainty 0.9]);
```

- An inference system that chooses a rule to use at each point in the consultation.

- An explanation facility that can tell the user why it has made a particular decision, and how it has reached its conclusion. This is essential if the system is to communicate on human terms. It is not enough for an expert system simply to offer a conclusion; the user, normally a skilled professional, will want some justification, particularly if he or she disagrees with its decision. At its simplest, the explanation facility can consist of citing the rules (in their English form) that are called on (or 'fired', to use the KBS term) during the consultation:

```
Is the main dish fish meat, meat, or poultry?
   > meat
What type of meat: pork, veal, lamb, etc
   > why
We are trying to determine the contents of the main dish
It has already been determined that
   the main dish contains meat
If it is also true that the main dish does not contain veal
Then we can conclude that there is strong evidence
   (certainty 0.9) that the wine colour is red
(by the rule colour1)
```

Rarely, however, will this correspond to the user's line of thought. Where the expert system has been following a line of reasoning forward from the given information towards a conclusion, the human user may have been reasoning backward from a tentative conclusion to the facts that might support it. Or the human may use some completely different method of reasoning, such as calling on analogous problems, or interpreting a mental image. The user may also want to know not only the rules that have been called on to support a conclusion, but also those that have not been used, or might have been used had the circumstances been slightly different. Providing good explanation facilities is one of the most difficult problems of expert system design.

1.4.1   Reasoning with uncertainty

An important characteristic of expert systems, one which sets them apart from most other computer programs, is the ability to deal with uncertainty, both from unreliable data, and from uncertain knowledge.

Data can be unreliable for a number of reasons: they may be 'contaminated' (for example data from laboratory tests for bacteria may be inaccurate due to stray organisms), they may be incomplete (because the user was unable or unwilling to answer a question) or they may be just plain wrong (because the user gave inaccurate information, or an automatic recording device was faulty). Two ways to deal with unreliable data are to make a 'best guess' — to compare a range of data with stored models, for example of typical diseases and their symptoms — and to weight the information according to importance: some data must be known and accurate in order to reach a decision, other data are merely useful as confirmation.

Uncertain knowledge may be due to doubt on the part of the human expert (for instance as to which wine is best with Veal Cordon Bleu) or to uncertainty of events (for example in a horse betting expert system, as to which horse will win the race). Statistical methods, such as Bayes Theorem, have been employed to handle such uncertainties, but the simplest method (used in many commercial and academic expert systems) is the one used in the example above: to add a numeric 'certainty factor' which indicates the strength of a conclusion. These certainty factors are then combined arithmetically to produce a final conclusion. Certainty factors have been criticised for being *ad hoc*; the main justification for them is that they 'seem to work' and that in any case they are not particularly sensitive: in practice, altering the certainty factors, even by large amounts, does not often greatly affect the final conclusion.

Expert systems in commercial use range from ones that are no more sophisticated than our Wine Advisor, with perhaps 50 rules and little or no explanation facility, to ones such as XCON (also known as R1), which has over 3000 rules. [6] XCON is in regular use by Digital Equipment Corporation to configure its computer systems and by the end of 1984 it had dealt with over 80 000 customer orders. Its expertise has been built up gradually since 1980, when it first saw commercial use, and now fewer than one in 1 000 orders is misconfigured due to lack of rule knowledge. There are still errors in around 10 per cent of its conclusions, mainly due to the fact that it has descriptions for only 5 500 of the more than 100 000 computer parts that could appear on a customer order, but most of the errors are minor and its conclusions can usually be used, though often only after a little human editing.

As people come to realise the limitations of expert systems, and the need for a human check on their output and reasoning, so effort has been devoted to making the systems easy to use and interrogate, with graphical displays, diagrams showing the order in which the rules were called, and different modes of explanation.

1.5   How expert systems are designed

The traditional method of designing a computer program is to create a detailed specification of the task, in some formal notation such as a flow diagram or a decision table, and then to write a program that corresponds directly to the formalism. The method assumes that the problem can be fully specified before a line of the program has been written and run. However, this is rarely the case in designing expert systems.

Generally, the greater a person's expertise, the more difficult it is to extract their knowledge in a form that can be encoded as a set of rules. Take a domain in which we are all experts, the spelling of English words. Given nonsense verbs such as 'plip' 'pleep', 'bolly' 'foy', you can easily form their past tenses: 'plipped', 'pleeped', 'bollied' 'foyed'. For each of these words you call on a different rule (such as 'if the word ends in a 'y' preceded by a consonant, then take off the 'y' and add 'ied' ') but it would take you considerable time to describe the precise rules for past tense formation, and the circumstances in which each rule is used. Although a human expert may behave 'as if' governed by rules, it is no easy job for the person designing the expert system (often called a 'knowledge engineer') to prise them out, and it is much more difficult if the knowledge contains uncertainties.

The standard design methodology is for a knowledge engineer (someone who understands both the design of expert systems and the psychology of eliciting knowledge) to begin by holding informal interviews with one or more experts in the domain, in order to gain a general understanding of the task and techniques. Then he or she will focus on aspects of the expert's knowledge, perhaps through case studies, and produce an initial design for representing that knowledge to the system. Often at this stage a very simple expert system will be knocked together, in order to test out ideas, and to get reactions from the experts. Gradually the knowledge engineer will produce an analysis of the task, a detailed representation of the domain and its knowledge, and a set of heuristics ('rules of thumb') used by the experts. These will form the knowledge-base and rules of the expert system. Next a prototype system will be designed and, in close consultation with the expert, the knowledge engineer will look for gaps, mistakes, and inconsistencies in the knowledge. Then the system will be tested on real data, its conclusions compared with those of the experts, and fine tuning carried out before it is packaged and sold.

This is by no means the only method of system design. The knowledge engineer and the expert may be the same person, either because a programmer has particular expertise or, more likely, because an expert in some field decides to try his hand at expert system design. This approach is likely to become more common in the future as software to help in the design of expert systems is produced. The simplest of these design tools is the 'expert system shell', which is essentially an expert system with the knowledge-base and rules removed, leaving the rule interpreter and explanation facility, to which new rules and facts, in a different domain, can be added.

In the early days of research in expert systems it was predicted that such shells could be adapted to a wide range of problems, by doing little more than pouring in new facts and rules. In fact, domains that are superficially similar may require very different types of reasoning and forms of explanation. For example, diagnosing diseases seems, at first sight, to be a similar process to diagnosing faults in a car engine. In some cases the techniques *are* similar — the human expert looks for a collection of symptoms that indicate a particular problem — but the procedures for checking and repairing the damage may be very different: a car mechanic has a complete circuit diagram of the car's wiring, but the doctor can only try a few probes, such as taking an X-ray, to discover a human's 'wiring'; a car mechanic can check a fault by replacing a part, but a doctor has to rely on measurement and more general intervention such as administering drugs. Despite the difficulties of providing all-purpose shells, as the tools become both more powerful and easier to use, expert system design may become a cottage industry, with 'home wine advisors', 'home repair guides', and 'tax advice systems' being developed by individuals on their personal computers.

Another prediction, still in currency, is that the knowledge engineer may soon be made redundant, as techniques are developed for expert systems to learn automatically from example, in the same way that a person acquires knowledge by observing and interacting with the world. Unfortunately, psychologists still have no detailed theory of how people learn new skills and concepts; what they do know is that learning is a constructive process. Knowledge is not just poured into the mind; new data must be translated into mental representations, from which knowledge can be extracted, and then integrated with existing skills or ideas. This is a complex process, at the frontier of research in artificial intelligence.

Systems that can learn new facts and concepts (of a logically simple kind) are already commercially available, but they are severely limited in ability to explain their reasoning. When people design expert systems, they attempt to organise the rules in a way that allows them not only to make appropriate inferences, but also to provide explanations to the human user. Rules learned automatically from examples are not so perspicuous.

In general, the ease with which a KBS can be designed depends largely on the accessibility of the knowledge, and the ease with which it can be expressed in one of the standard knowledge representation formalisms. The latter problem will certainly recede as new techniques for knowledge representation are developed, but the difficulty of acquiring knowledge from human experts is likely to remain, at least until powerful and general rule learning programs are developed. Acquiring knowledge for a KBS is more like unravelling a large tangled ball of string than sucking up fluff with a vacuum cleaner.

1.6   Other types of KBS

Expert systems are not the only type of KBS to be commercially available now or in the near future. Computer vision systems are coming into use, generally in combination with a robot for recognising and manipulating objects. A typical application might be a warehouse robot that can recognise and stack crates of a standard size and shape. Planning systems start with a small set of goals and specifications, such as 'plan an itinerary for a one week camping holiday in the South of Spain using recognised camp sites, taking in historic sites, and avoiding the major tourist spots', and produce a plan, in the form of a set of operations that can be either carried out by an automated system such as an industrial robot, or (as in the example here) reported to the human user.

So far, the assumption has been that each KBS stands alone, interacting with one human user, and containing knowledge derived from a single expert (or a small and identifiable number of them). But there are many other possible modes of interaction. A KBS might serve many users, in a large office or through a network of terminals in, say, travel agencies or banks. Or the systems themselves may be linked together, one system querying another, first to assemble information (say about a person's credit transactions with a number of agencies) and then to produce a conclusion (about that person's general creditworthiness). Or the knowledge in the system may be gleaned from many sources. Television service engineers of the future might carry KBSs out into the field to advise them on fault repairs. If they encounter a new type of fault, then they might have the facility to alter the knowledge-base, or to add new rules to cover the problem. A variation of this is a public knowledge-base which people may call on either to use or to store knowledge. A medical practice, for example, may in the future store not only patients' records on a computer, but also store rules for treatment and for drug prescription.

1.7   The future of KBSs

In 1987 the US Office of Technology Assessment[7] carried out a survey of leading AI researchers, asking them to forecast the future state of the field. The majority of them asserted that there are 'few fundamental barriers for machine intelligence and thus that over the long term intelligent machines would meet or surpass humans in most cognitive skills'. For the short term (the next twenty to thirty years) their views were mixed, ranging from the 'minimum progress' scenario that expert systems will be widely deployed as decision aids, but will be extremely limited in subject area, to the 'maximum progress' one that 'intelligent machines will be a near-constant companion to most humans, resulting in a new level of intelligence through human-machine symbiosis'. The summary of the 'moderate progress' view is worth quoting in full:

> Expert systems will be somewhat restricted in subject area but extremely powerful and in some cases will be given *de facto* control over routine decisions. For instance, an expert system (or network of expert systems) might essentially have mastery over most of the field of medicine. It would learn by reading articles. It would not deal with ethical issues. Expert systems would substantially augment individuals' knowledge and access to knowledge.

One of the problems with forecasts of the future of technology is that they almost invariably omit some crucial social or political factor which alters the entire course of research and development. A thirst for computer-provided knowledge, or conversely a rejection of the 'information society', a move towards political or economic decentralisation, or increasing State or monopoly capital control, would all have a profound effect on the development and deployment of knowledge based systems.

In the 1960s many predictions were made about the public access to computers. Mostly they assumed that large public databases would be set up in towns and cities, to which the public would be connected through terminals in schools, workplaces, or the home. What none of them foresaw was the rapid growth in personal computing, with general purpose microcomputers being sold for less than £100. There may be a similar boom in small-scale personal KBSs, for use in the home or office. If so, those directly affected will include people whose advice and decisions affect daily life in an intimate fashion: doctors, lawyers, teachers, town-planners, civil servants, policy makers in national and local government, not to mention the military. No profession is in principle excluded. Even someone who does not use an expert system in the home will be *indirectly* affected, because of their use in the professional institutions through which society is organised.

References

1. *Alvey Programme Annual Report 1986.* Alvey Directorate, Millbank Tower, Millbank, London.

2. Dray, J.A. (1982). Coping with the Thinking Machine. Unpublished thesis, Wesleyan University. Quoted in *The encyclopedia of AI*, (eds.

S.C. Shapiro and D. Eckroth), p. 1058 (1987). Wiley, New York.

3. *The Times*, November 1, 1946.

4. Nilsson, N. (1984). Artificial intelligence, employment, and income. In *The AI Magazine*, Vol. 5, No. 2, pp. 5–14.

5. Hayes-Roth, F. (1984). Knowledge-based expert systems — the state of the art in the US. In *Expert systems: state of the art report*, (ed. J. Fox). Pergamon Infotech, Oxford.

6. Bachant, J and McDermott, J. (1984). R1 revisited: four years in the trenches. In *The AI Magazine* Vol. 5 No. 3, pp. 21–32.

7. Office of Technology Assessment, US Congress (1987). Artificial intelligence: a background paper. US Government Printing Office, Washington, DC. Cited in *The encyclopedia of AI*, pp. 1050–1.

2 Applications of KBSs

So far, those KBSs that exist outside the research laboratory have been mainly employed to support decision making in organisations that are both rich enough to invest in new technology and innovative enough to consider new solutions to the problems which demand human expertise. But once the technology becomes cheaper and more reliable then it will be routinely used by the more cautious companies, by government agencies, by welfare organisations and by individuals. It will be routinely used to provide more cost-effective ways of managing, for example, production, stock control, personnel, or marketing, and it will be retailed as a successor or addition to home microcomputers.

The term KBS or expert system may still remain, to cover a general class of systems, but just as the field of computing has now been divided up — by function (hardware, software, peripherals), by application (data processing, number crunching, word processing), by user (programmer, manager, secretary) and by size (micro, mini, mainframe) — so too will the territory now held by KBSs. The divisions may well be the same as for computing in general — function (advice giving, planning, assistance, tutoring, monitoring, control), application (health, finance, law, business management, production management, military), user (programmer, knowledge engineer, manager, home user) or size (there are no established categories for the size of a KBS; the number of rules is one guide, but only for systems that are primarily rule-based). In this chapter we have organised the sections according to application. Within each section, and within the chapter as a whole, we hope to indicate the range of functions and users of present-day KBSs.

2.1   Finance

Financial problems, and a need for relevant expert advice, face every one of us. Individuals make loan applications, tax returns, plans for retirement, claims on insurance, and investments of various kinds; and some have to brave the thicket of D.H.S.S. regulations to find solutions to their money problems. Managers must do all this on behalf of their companies, and in addition deal with company law, forward planning for tax efficiency and company development, financial aspects of employment law, and sometimes mergers and take-overs too.

Financial problems have four aspects: arithmetic, rules, priorities, and plans. The arithmetic can easily be done by computers. The rules include many that are clearly suitable for programming in KBSs, and others which can be usefully automated also. The priorities are more elusive, for it may be difficult both to identify them and to order them in importance; even so, there is significant scope for clarifying the interconnected constraints that face people making financial decisions. Finally, planning of certain kinds can already be done, or at least made easier, by KBSs.

The rules involved in financial decision-making are sometimes explicit and sometimes implicit, sometimes without any exceptions and sometimes mere 'rules of thumb' that can be ignored on occasion. As in other domains, a KBS can be produced for financial applications only to the extent that these rules can be made explicit. Some financial rules are not only published, but expressed in a reasonably clear fashion; these are clear candidates for embodiment in a KBS. Examples include aspects of tax law, company law, and government regulations covering applications for loans and grants. Other rules are known only to a select group, and not all of them are written down. Examples of these include the practice of auditing; the conventions a bank-manager uses in deciding whether to offer a mortgage, or a loan for a car or business venture; and the considerations of a stockbroker when playing the market.

The priorities, or values, relevant to financial decisions are very diverse. But some crop up over and over again. In planning for retirement, for instance, people commonly have to decide whether they want to leave a nest-egg to their children, whether they will need extra income, whether they intend to sell their house and/or buy a cottage, whether they will need cash for spending (on travel, perhaps) or are happy to have their money locked into a pension fund, and what degree of risk they are prepared to accept in investments. In corporate planning, a manager's 'bottom line' may be to make forty people redundant; but whether he or she chooses to do so, and if so which employees are picked, will depend on factors like employment law, the general priorities guiding personnel management in the company, and the loyalties (and prejudices) specific to particular individuals. Such constraints and 'subjective' values, together with their differential importance and their trade-offs, can be included among the rules in a KBS. The person making the decision can thus be reminded of these priorities, and advised on whether and how they can be achieved.

2.1.1   Existing systems

The financial KBSs that are already being used are not all available to the general public, or even to other professionals. Bankers, accountants and the like sometimes rely on 'in-house' written manuals, which represent the accumulated expertise of the firm over the years. Such manuals can now be used as the basis of an in-house KBS (which, for obvious reasons, is not circulated outside the institution). As a result, KBSs in the financial world are often custom-built, either by the employees of the professional institution itself or by a computer consultancy firm on its behalf.

The Personal Financial Planning System (PFPS) was developed for a large US bank and is the central resource for a new division of the bank that provides personal financial services to families. The service is offered for a few hundred dollars per consultation, and the bank estimates that to provide similar quality services with conventional methods would cost over $10 000 each time. The KBS took several years to develop with a team of up to 10 people.

Syntelligence's Underwriting Advisor has been installed in several insurance companies. The basic system provides standard underwriting functions, and then requires considerable modification to fit the requirements of the individual company.

ExperTax is a KBS to support tax planning for tax accountants within Coopers and Lybrand and is in use in all the company's 95 offices, running on IBM PC computers.

Changes in the central policy of an institution can be put into practice immediately, and effectively, by distributing identical newly-written KBSs to all the subsidiaries. The Caisse d'Epargne Bank has supplied its branch offices with a KBS for advising the local staff on whether to grant home loans. And many banks (and other moneylenders) now use in-house KBSs to assess the credit-worthiness of individuals seeking loans for various purposes.

A system currently under development is intended for use by the staff and clients of the D.H.S.S. This large-scale, five-year, Alvey Demonstrator Project was outlined in Chapter 1. The system will provide guidance to claimants and clerks (and, in principle, to Citizen's Advice Bureaux) on the complexities of the regulations for welfare benefits of various kinds. At present, many people do not claim their full entitlement, because neither they nor the clerks in their local D.H.S.S. office are aware that they are eligible for certain payments. Others may know they are entitled to a benefit, but have difficulty (even with the help of a D.H.S.S. clerk) filling in the application forms because they are so complicated. Putting the regulations into a KBS can not only clarify the rules themselves, but also in effect provide an application form that is tailor-made to the individual claimant. Whereas a paper form will often include many questions that the claimant is instructed to ignore, an equivalent KBS would enable the claimant to focus on the relevant points without even being aware of the questions that do not apply in the individual case.

A significant number of small KBSs running on personal computers is already commercially available. Some are aimed at the general public. But most are intended for use by specialists in giving financial advice and/or making financial decisions: accountants, bankers, insurance managers, stockbrokers, and moneylenders of various types.

2.1.2   Computerised stock dealing

Some financial systems under development are designed not to respond to specific queries on demand, but to be left running in the background and to alert the user if and when a situation of interest arises. For instance, a commercial software company is producing a system that will monitor the fluctuations in the exchange rate of various currencies and international interest rates. Using its complex currency-dealing rules, it will continually test the combination of these factors to see whether they provide an opportunity to make a profitable deal. If they do, it will warn its user accordingly. In cases like this, the speed of the KBS's decision-making gives it an important advantage; once one currency dealer has a functional system, the others will be forced to follow.

Conventional computer programs are already used for trading in stocks and shares, programmed to trigger *buy* or *sell* orders when the software detects certain conditions or trends. A report by Michael Gemignani [1] points out the consquences of such computerised dealing:

> This so-called *program trading* has been suspected of causing wide swings in prices over a short period of time ... [The] securities market has become so complex, large, and fast-paced that it cannot safely allow humans to make key decisions ... [If] the traders' computers use software that makes essentially the same decisions from the same available data, then even greater market volatility can be expected. A worst-case scenario would be where all of the programs simultaneously sent signals to sell and none sent *buy* orders. The entire market would sink like a rock.

A slump caused by program trading would not be long-lasting, since traders would realise that prices were artificially low and begin to buy, but the result of program trading seems to be a greater volatility and unpredictability in the market.

These instabilities are not likely to diminish in the future. The argument runs as follows. Big gains can be made on the securities markets by the short term buying and selling of 'stock index futures', paper contracts representing the future value of a bundle of stocks.

> Stock index futures turn stocks into a commodity like silver or wheat. When the commodity future consists of a bundle of hundreds of stocks, each of which is traded on the open market, one needs a computer to undertake the complex analyses required for *buy* or *sell* decisions.

So, on top of the cycles of economic activity, lasting months or years and affecting stocks and shares, there are rapid fluctuations in the value of 'futures'. There is great advantage to financial dealers in being able to ride these waves, selling and buying from minute to minute as prices rise and fall. The movements are so rapid and complex that it is already necessary to use computers to spot the trends. The organisations with the better programmed computers will gain most. As trading programs become more sophisticated they will detect more and more subtle trends. Some of these trends will not be generated by real economic movements, but by the buying and selling activities of other computer programs. So, in the future, computerised trading may have little to do with economic realities, but become a gigantic computer game, program pitted against program, with the winners being those best able to spot any short-term trends, however caused. The result could be a fast-moving money market that is decoupled from the longer economic trends, and with only a few companies, owning powerful computers, able to play it succesfully.

2.1.3   Money-saving KBSs

Some KBSs used by financial institutions do not deal with 'financial' problems as such, although they can save the company a lot of money. Thus a major credit-card company is able to reactivate its computer quickly in the event of a breakdown by consulting a KBS running on another computer. The program is an automated computer manual: instead of having to leaf through pages of text to find the relevant instructions, the operator is asked an appropriate sequence of questions and so guided as directly as possible to the correct actions. Speed is crucial here because the company could lose a great deal of money if fraudulent card-transactions could not be checked.

2.1.4   Implications

That KBSs for solving money-related problems have entered the financial
world is certain. What is less certain is whether this is a good thing. For
instance, there is a fear that widespread use of computerised currency and
stock dealing systems could threaten the equilibrium of the world's markets;
but some financial experts believe this fear to be unjustified. Again, some
people fear that an automated 'lending manual' would inhibit a bank's employees from exercising their individual judgement in risky cases, not least
because they might be less experienced than those bank personnel who are
responsible for granting loans today. Certainly, time and aggravation would
be saved, because the local staff would not need to seek approval from head
office except in non-routine cases. And their decisions would be 'fair', being
governed by identical criteria and free of individual hunch or prejudice. But
some hunches are well-grounded. The less the branch officer's judgement
was exercised, the less scope there would be for imaginative lending.

As in all other domains where KBSs will be used, the financial firms and
institutions that use KBSs should develop procedures providing a safety-net
to catch 'maverick' cases that require the judgement of a human specialist.

2.2   The law

A KBS dealing with the law would not merely search a legal database for
case information using 'key-words', as current programs do. If it were used
to search at all, it would do so in a more flexible and intelligent way: seeking
cases having a similar structure, not labelled with the same key-word. Or
it might represent some part of the law in a way that could be tailored to
a specific individual, perhaps giving advice on what the legal possibilities
in a given situation are and how they can be pursued.

2.2.1   Potential users

The users of KBSs in legal contexts potentially include many different people: professional lawyers (solicitors and barristers) advising their clients;
the staff of Citizen's Advice Bureaux; Government Departments drafting
new laws or regulations; law students, or novice professionals, unfamiliar
with a certain area of the law; and judges and magistrates. Quite likely,
shops and public libraries will soon be offering programmed manuals on
do-it-yourself conveyancing, or how to make a will. But there are pitfalls
even in the law of conveyancing, which is a relatively straightforward area;
and home-made wills are at the mercy not only of the intricacies of the
law of probate but the ambiguities of the English language. Reliance on a

legal KBS (as on KBSs in other specialist domains) without the benefit of professional advice could often be misleading.

For those who cannot afford such advice, or who experience the average law-office as intimidating, the Citizen's Advice Bureau offers an alternative. There may be no legally qualified person available, as there would be in the (increasingly under-funded) Law Centres, but the staff are experienced in dealing with the commoner sorts of legal/bureaucratic problem. In the great majority of cases, deciding on the applicant's best course of action is trivial compared with identifying their problem in the first place; this is especially so if the applicant is relatively ill-educated and inarticulate. Suitable legal KBSs could help CAB staff to discover and analyse the relevant facts, before giving advice on what to do. Some systems might also be able to provide a specimen letter, either wholly 'canned' or constructed by adapting a skeleton letter schema in the light of the applicant's answers to certain questions. The letter (dealing with the law of tenancy, taxation, or employment, or with Council regulations on fire protection or pest extermination) could then be sent by the applicant to the relevant official or department. The system currently being developed (under the Alvey initiative) to give advice on D.H.S.S. regulations and entitlements could also be helpful in this context.

We have also been told that JUSTICE (the British section of the International Commission of Jurists) is currently developing an expert system which both lawyers and voluntary advice agencies can use to interpret and apply the Rehabilitation of Offenders Act 1974, a statute so densely drafted that very few of those who were intended to benefit from it ever do.

2.2.2 Applications of legal knowledge

Suppose that professional legal help is sought: what then? In order to be able to advise a client, and to take the necessary steps to ensure that his legal rights are protected, a practising lawyer needs to have and apply expert knowledge of at least four different kinds.

First, the lawyer must diagnose the client's problem, so as to determine the legal category into which it falls. For example, is this a question of the general law of contract, or of landlord and tenant, or of employment? Next, the lawyer needs to ascertain the particular rules of law which apply to the given case. These may be contained in legislation (a statute, or official regulations), or in rules laid down by previously decided cases, or both (for instance, a case which interprets a particular phrase in a statute). Third, the lawyer must decide what would be the best tactics to pursue on his client's behalf: to write a threatening letter, or first await further developments; to institute legal proceedings and, if so, in which court; to offer at some stage to settle the dispute, or to wait for the other side to come forward with such a proposal; and so on? Last, the lawyer must offer a prognosis of what the final outcome of the dispute is likely to be (and what is the worst-case alternative).

2.2.3 Identifying the client's problem

The first of these tasks (identifying the client's problem) is the one for which practising lawyers have the least to gain from KBS-aids. To experienced solicitors, the answer in most cases is obvious — so no KBS is needed. (Although we noted above that less well-qualified counsellors, such as the staff of a Citizen's Advice Bureau, might find KBSs helpful in diagnosing legal problems.) For the rare occasions where the legal diagnosis is not obvious to an experienced lawyer, it would be almost impossible to devise any helpful rules — so no relevant KBS could be produced.

2.2.4 Determining the applicable laws

The second task (determining the applicable laws) lends itself better to the development of expert systems. This is especially true where the concern is with statute law. It is the explicit objective of the lawyers who draft Acts of Parliament, as well as the judges who interpret them, to create systems of rules that are precisely specified, logically complete, and coherent. So it should be possible to provide useful KBSs for areas of the law governed by statute: the JUSTICE system mentioned above is a good example.

Another example of a legal KBS is a program that represents the British Nationality Act of 1983. The program (an Alvey-funded project, not available to the public) is written in a programming language modelled on formal logic. And the Act contains a number of rules that can be expressed in logical form, such as:

> IF the client was born in the United Kingdom
> AND at least one of his parents was settled
>   in the United Kingdom at that time
> THEN the client is a British subject

To apply this criterion, a lawyer (or immigration officer) needs to know what counts as 'settled'. Another section of the same Act provides rules defining what is meant by this term, and these interpretative rules too are contained in the KBS.

Notoriously, the interpretation of a word or phrase in statute law often turns out to be less straightforward than the legislators had assumed. So the courts have to refine the interpretation of the terminology continually, as the cases brought before them discover the unclarities. These decisions, also, can be embodied in a statute-KBS (if it is designed to allow the addition of new rules). However, the attempt to draft a statute into KBS form can itself highlight unsuspected ambiguities, and prompt the draftsmen to clarify the wording. Indeed, if the statute were to be drafted in the first place with the help of KBS technology, the unclarities would very likely be significantly reduced. (A state legislature in the USA has for several years been drafting its regulations in programmed form for this reason.)

Statute law, however, is not enough. There are still many areas of the law of England and Wales which are not governed by statute, but have been created largely through the precedents of cases already decided in the courts. Sometimes, it is possible to express 'rules' reflecting at least part of the relevant case-law. One example of this is the law of negligence, which is governed by such rules as the following:

IF the client owed another person a duty of care
AND the client acted in breach of that duty
AND the other person suffered damage as a result
THEN the client is bound to compensate the other
    person for that damage

IF the other person is someone whom a reasonable
    person in the client's position should have
    had in contemplation as being likely to be
    affected by the client's acts
THEN the client owes that other person a duty of care

As is only too clear from these illustrations, however, rules of this kind tend to be phrased more generally than statutory rules are. The meaning of words like 'reasonable', 'duty', or 'damage' (not to mention 'likely to be affected by') cannot be precisely defined by adding a few interpretative rules. One might even argue that to capture them satisfactorily one needs to make explicit various underlying principles of jurisprudence — an exercise that is not impossible in computational terms, but which requires coherent decisions to be made about fundamental legal questions. Expert systems incorporating the law of negligence or similar laws would therefore tend to be larger, and less reliable, than those reflecting statutes.

Nevertheless, future KBSs might help lawyers to identify the laws applicable to their clients' situation even if case-law (not statutes) were in question. The KBSs concerned would have to be very different from to-day's systems, because they would have to embody analogical reasoning.

Research on analogical thinking, and its underlying computational principles, is being done by a number of workers around the world. Some of
them are specifically interested in the use of analogy by lawyers, in how
one legal case can be recognized as being significantly similar to another
case superficially different from it. Such research might be applied in future to build legal KBSs capable of noticing useful analogies between the
client's situation and past cases. Even so, it is questionable whether any
but the very largest (and most expensive) would be of much practical use
to experienced lawyers. (As with other specialisms, analogical KBSs might
be useful to law students as teaching aids.)

2.2.5   Choice of tactics

The third area of legal expertise (the choice of tactics) might be amenable in
part to the development of KBSs. Here, what needs to be brought together
is not so much a set of rules explicitly designed to be comprehensive and
consistent, but rather the accumulated experience of practitioners in legal
affairs. Many of the tactical heuristics, or rules of thumb, followed by
lawyers are based on their general experience of human beings. Examples
might include:

> IF your client's case is weak
> THEN press the other side as hard as possible
> AND do not initiate any offer of settlement
>
> IF you know that the other side is poorer than your client
> THEN make the litigation as expensive as possible for them
>
> IF your client's case is so bad that he or she will have to pay
>   in the end
> THEN spin things out as long as possible

The rationale for these practices lies in general facts about human psychology; the first, for instance, presupposes that this behaviour will tend
to make the other side think that your client's case is stronger than it really is. But individual litigants are different: what will be accepted by one
opponent as reasonable behaviour, or at least as fair play, will be seen by
another as a red rag to a bull. It is not clear that any practical benefit
could be provided to experienced lawyers (as opposed to trainees) by KBSs
designed to represent this sort of expertise.

Tactical expertise in technical areas of the law is another matter. It
may be that the best way of making progress with a particular sort of legal

task can be faithfully, and usefully, expressed in a KBS. Conveyancing, for instance, or setting up a company or partnership of a certain kind, can involve complex procedures whose nature — and order — could usefully be identified in a KBS.

2.2.6 Prognosis of the client's case

The fourth task facing a practising lawyer (the prognosis of the client's case) demands judgement and experience of many different kinds. Some is specific to certain individuals. Thus a lawyer may predict that if the client is arraigned for rape before a particular judge (a situation the lawyer will try to prevent) and found guilty, he will receive a very severe sentence. Or the lawyer may (whether he or she admits it or not) base the prognosis largely on a judgement about the likely effect on the judge or jury of the client's less endearing personal characteristics. It is hard to see how a KBS could be designed to take account of such idiosyncracies and subtleties.

But some other legal prognoses are based on judgements of a type which could usefully be computerised. One example concerns the assessment of damages for personal injuries. When someone is injured in an accident, for instance on the roads or at work, and can establish that the accident was due to the 'fault' of someone else who owed him or her a duty of care, the injured person may bring an action for damages to compensate for the injuries suffered. Assuming that it is successful, the amount of the damages will be calculated according to what are now very well-established principles. Among the many factors to be taken into account are the victim's age, past earnings, future prospects with or without the injuries, extent of disability, likelihood of improvement or deterioration of the condition, actuarial considerations, inflation, etc.

A well-known text-book (*McGregor on Damages*) contains reports of a very large number of decided cases, and is kept regularly up to date. Since the rules are known with an unusual degree of precision, and since there exists a substantial published database, it should be comparatively easy to construct a KBS in this domain.

A program for assessing damages for personal injuries could be used by a lawyer to advise his client on the likely size of the award were the case to go to court, and to negotiate an out-of-court settlement with the other side's lawyers. If the lawyers on both sides were to use the same KBS, they ought to have no difficulty in agreeing the asessment of damages, and the only remaining scope for disagreement would be on the attribution of 'fault'. Indeed, a KBS for the assessment of damages could be helpful to judges too.

This is not the only example of a feasible KBS that could be of use to the judiciary. Judges and magistrates about to pass sentence might use a KBS to recall what sentences had previously been pronounced for similar offences; to inform them of the effects of those sentences on different types of culprits (if this is known); to remind them of the range of possible sentencing options; and to embody various aspects of sentencing policy for their guidance.

When someone is found guilty of a criminal offence, either because he or she pleads guilty or on a verdict after a trial, there is usually a wide range of 'disposals' available to the sentencing court: absolute discharge, conditional discharge, probation, community service order, fine, imprisonment, committal to a mental hospital for treatment, etc.

Until quite recently, magistrates and judges chose between these different disposals largely 'by feel' or 'in the light of experience', untroubled by any independent research on the effects of different disposals on different offenders. As a result, there were wide discrepancies between the policies persued by different courts. Moreover, it has never been — and still is not — very clear what objectives courts think they pursue when they sentence offenders, as between punishment, deterrence, reform, rehabilitation, treatment, prevention, retribution, deserts, reparation, restitution, or correction. In recent years, this situation has begun to change. The judges now have regular sentencing conferences among themselves and the Home Office has begun to conduct research on the effect of different disposals on different categories of offender.

As more research in this area is carried out and more sophisticated results are reported, differential effects of the various disposals may be identified. In that case there might be scope for a KBS that could be consulted by the judiciary before passing sentence. One day, perhaps, we may hear a Crown Court judge address the man in the dock on something like the following lines: 'Norman Stanley Fletcher, you have pleaded guilty in this court to a series of offences. In the light of your previous record, my normal inclination would be to put you on probation. However, I have consulted DISPOSAL, the expert system which is running on the microcomputer I have on the Bench with me. Having entered all the relevant particulars about you, it tells me that the prospects of your responding favourably to probation are only in the range of 14 to 17 per cent, and that the statistical significance of this result is very high. It also appears that the prospects of success of any other non-custodial sentence that I might consider would be even lower. Accordingly, I have no option but to sentence you to a term of imprisonment of seven years — not with any hope of reforming you, but simply in order to protect the public from your depredations while you are inside.'

Whether a successful appeal in such a case could be based on the judge's admission that he over-ruled his own judgement because of what his micro-computer 'told him' is an intriguing question.

2.3 Education

2.3.1 Automated teachers

The idea of an automated teacher is a seductive one. Instead of giving a child 1/30th of the time of a human teacher — one who may not be trained in the subject he or she is teaching and who works with out-of-date and inadequate materials — each child could be offered a personal auto-mated tutor, equipped with the latest knowledge, able to answer compli-cated mathematical questions, able to call on a vast source of information, and never subject to fatigue, or lapses of concentration. The human teacher would be promoted to classroom manager, responsible for preparing course work and tending a flock of teaching machines:

> For so long teaching has been regarded as a human task that it is novel to suggest a machine should take over the role of contact with students, and leave a teacher to do the planning and preparation of the lesson. But it does seem to work, and in a world that is short of teachers there is every reason to develop it as far as possible. [2]

If the computer can act as a vast store of knowledge, then why not transmit it in neatly packaged chunks to pupils? This was the prevailing vision of computers in education during the 1950s and 60s. Computer systems of this type were developed, and some are still in use, providing training in, for example, arithmetic, geography and foreign languages. The reason why they have never transformed the classroom is that they are restricted to transmitting and posing multiple-choice questions. They are not 'automated teachers' so much as 'electronic page turners', giving out chunks of prepackaged material, in predetermined sequences.

The transmission of information is only one method of teaching. A trained teacher will build on a child's existing knowledge and suit the style of teaching to the learner's needs and difficulties: discussing the context of the lesson, identifying weaknesses, offering examples and remedial exercises to the slow learners and additional project work to capable ones. If a KBS is to be effective as a teacher it must display a similar range of skills, founded on a knowledge of the subject, the child and methods of teaching. It must build up, or be given:

- knowledge of how to teach (which includes knowledge of students in general);

- knowledge of what is being taught;

- knowledge of who is being taught;

- knowledge of how to apply this in particular cases.

So a system to teach calculus should firstly be able to *do* calculus: to perform symbolic differentiation or integration on any example that a pupil presents. Secondly, it should be able to compare a student's solution with the correct one in order to identify whether, for example, it is the result of a careless error or of a genuine misconception. Thirdly, it should be able to offer advice and comment appropriate to that student's needs. Lastly, it should be able to offer teaching that is appropriate to the circumstances: an initial lesson, a tutorial, exercises, remedial work.

2.3.2 SOPHIE

To give an example of what is currently possible, SOPHIE, standing for SO-PHisticated Instructional Environment, is about the most complex single teaching KBS yet built. It is designed as a 'reactive learning environment' — one that encourages students to explore ideas and hypotheses in a particular domain — and it provides detailed feedback about their logical validity. The domain with which SOPHIE deals is that of electronic troubleshooting, teaching people the skills for finding and curing faults in some non-trivial piece of equipment. The subject of electronics is sufficiently well understood for it to be accessible to computer simulation, but educationally it is very elaborate; a domain in which there is no reasonable substitute for experience.

In operation, SOPHIE shows the student a circuit diagram, and chooses some fault that the student then has to find by requesting measurements of voltages, currents etc. in parts of the circuit. The dialogue is conducted in a subset of English. The students can float a hypothesis as to what is wrong, ask for a part to be replaced, or ask for help. In each case SOPHIE will analyse the suggestion or request in the light of what the student has learned so far, and generate a suitable response.

SOPHIE is a very large program but is remarkably efficient, with a response time of the order of three seconds. This, together with its very powerful inferencing capabilities, gives it the edge in some ways over a human teacher. Nevertheless, it is definitely a 'reactive learning environment' rather than a tutor, since:

- it has a very poor model of the student: nothing but the list of measurements that the student has requested;

- it has no teaching strategy (it cannot select particular faults that would be appropriate to the student's general level of skill or past mistakes);

- it cannot expound the theoretical knowledge that the subject requires. The student is assumed to have this beforehand, and SOPHIE does not look for the lack of it. It merely offers a very good environment in which to practise and hence increase troubleshooting skills.

For all its limitations SOPHIE is an impressive piece of work, and one which has, so far, taken 15 years to design. Although many of the *ideas* embodied in SOPHIE could be transferred over to other domains, or other styles of teaching, very little of the computer program itself could be reused.

A few other successful computer-based teaching and tutoring KBSs have been written but, in general, the history of computer-based training has been marked by frustration and disappointment. Designers have often grossly underestimated the time and effort needed to design the teaching material and many of the finished systems behave like didactic, uninformed, inflexible schoolmasters.

2.3.3 A generalised teaching system

So far there is no equivalent of an 'expert system shell' for teaching: a core system to which new knowledge can be easily added. There are two reasons for this: teaching has not attracted the reseach and development funds that have poured into expert systems, and tutoring is a highly demanding task: it requires the integration of general and domain-specific knowledge, an understanding of the learner's skills and misconceptions, and a range of strategies for presenting new concepts and skills, answering queries and assessing the learner's state of knowledge.

Designing systems that permit a teacher or educational designer with no programming skills to add knowledge and a teaching scheme for a particular subject is a formidable research problem. But such systems should be available within the next ten years and that is an exciting prospect, not least because it may encourage teachers and curriculum designers to think deeply about the purpose, content and structure of teaching, just as expert systems are at present encouraging doctors to rethink the process of medical diagnosis. Many educational issues will be brought into the open: should a computer-based economics tutor contain knowledge about a Keynesian economic system, a Marxist one, a monetarist one, or a blend of all three?

And what of an English teaching system, given as many prescriptions for literary style as there are authors?

2.3.4   Support systems

There are other approaches. A learner can be provided with 'knowledgeable support' during a demanding task, such as writing or mathematics. A support KBS has a limited teaching strategy — the learner directs the activity and the system acts as an assistant and advisor — but it reduces cognitive load by taking over demanding but low-level activities (e.g., in mathematics, performing the basic calculations) leaving the learner more free to concentrate on higher-level aspects of the task, such as formulating the problem and planning a solution. Such tools already exist to help computer programmers, and they are being adapted for educational use in areas such as writing, mathematics, and physics. For example, a project is being carried out at Sussex University with the aim of producing a *Writer's Assistant* that will support a writer in the production of complex documents, guiding him through the writing process from creating and organising ideas, to producing and laying out the final document.[3]

2.3.5   Software for special education

The use of computer systems to support people with disabilities or special needs is still at a relatively early stage, but shows great promise. Expertise in robotics has been used in the design of vehicles and manipulators for the disabled, research in human/computer interaction has shown ways of providing disabled people with access to computer tools such as word processors, and cognitive science research offers the promise of programs suited to peoples' cognitive abilites and limitations.

A scheme entitled COMET (Concerned Micros in Education and Training), administered by the National Bureau for Handicapped Students, is investigating ways in which people with limited muscular ability can control devices such as a powered arm. They are also developing new communication systems, for example a program that will convert the sounds of people with speech disabilities into more normal speech, and producing speech-to-text conversion systems for the deaf. These are long-term projects, but less ambitious ones include the design of document readers with speech output and programs that can adapt to disabled typists by, for example, offering abbreviations for commonly used words or phrases.

2.3.6   Exploratory environments

The most successful of all the contemporary educational software is the
Logo programming language. Although not a KBS, it was developed in Artificial Intelligence Departments at M.I.T. and Edinburgh University and
was inspired by research in both cognitive psychology and artificial intelligence. Logo is a language for learners, providing children with an easy
introduction to the computer through 'turtle geometry', software that allows a child to command a 'turtle' (a small motorised cart with a pen
attached) to draw shapes and patterns, first by giving it direct commands,
and later by writing programs to carry out more sophisticated operations.
Additions to Logo allow the child to explore other domains, such as the
physics of motion, or language, and the idea is that the child both learns
about the particular domain and gains general problem-solving skills. By
'teaching' the computer (writing a program to enable it to perform new sets
of activities) the child gains skills in organising ideas, planning, and testing
the results (the computer's actions) against the plan. Learning becomes
a playful activity, and mistakes in the programming are not errors to be
penalised but 'bugs' to be corrected (sometimes a 'bug' in a program may
even lead to a new, more interesting, activity, or to a greater understanding
of the problem).

2.3.7   The design of educational software

The main problem with introducing computers into the conventional classroom is the role of the teacher. In the early days of educational computing
there was a great deal of mutual animosity between teachers and the designers of teaching programs, with the designers claiming that their system
would transform education and free the teacher's time for individual tu-
ition, and the teacher seeing only a complex unreliable gadget, providing
inappropriate teaching by old-fashioned methods. More recently, teachers
have been encouraged to create their own teaching software. Some of this
has been imaginative and appropriate, but much of it has been no better
than the early systems.

   An advantage of 'home grown' software is that the author has a personal
commitment to its success and is willing to tolerate its limitations. The
difficulty is that a program which just about works under the supervision
of the teacher who designed it will generally be both inappropriate and
unreliable in any other hands. Teaching is the only profession which relies
on software produced by colleagues, rather than professional progammers.
As John Self[4] points out:

Practising teachers have a healthy scepticism about so-called 'research'. However, most of what is best about present educational computing is the product of many years' research, not the product of a cottage educational software industry. For example, Logo (one of the few redeeming features on a bleak landscape) was inspired by Piaget's life-long research on developmental psychology, was designed at a private research establishment in the late 1960s and was then fostered for several years in artifical intelligence research laboratories ... the subject of artificial intelligence remains the most promising source of new ideas for educational software. The fact that there are today so few products from research in evidence is not altogether the fault of researchers: support for such research over the last decade has been paltry. There is very little research now being carried out to complement the expenditure on educational hardware, and almost no longer-term research.

One successful approach to the design of educational software has been 'team design': a group of people, including a programmer and a teacher (and usually others, such as an interface designer and a curriculum expert) work together on the producing the system. Each person has special responsibilities, but all are sufficiently conversant with the project, and with the other tasks, that they can offer advice and criticism. Both the Open University and the University of California at Irvine have used this approach to create innovative and useful software.

2.3.8   The short term

In the short term the most successful applications of KBSs to education are likely to be in training adults in specific skills, and supplementing the knowledge of professionals such as doctors or engineers. An example of a skills tutor is CLORIS, a prototype video-based teaching system, which can display a training film, such as one to teach the skills of reading micrometers (a type of scientific measuring instrument). At any point the student can interrupt the film and enter into a discussion about its content, asking questions about the event in progress ('Why was ... done before ... ?', 'Tell me more about ... ') or the objects on the screen ('What is ... for', 'Show me a better view of ... '). Other current projects include the design of tutors for cardiac radiologists, boiler engineers, submarine commanders and oil platform operations technicians. For such applications, the learners are likely to be highly motivated, at a similar level of ability, and without the severe misconceptions or learning difficulties of some children. Thus the system designers can concentrate on representing subject knowledge and avoid the difficult problems of modelling the learner's knowledge or providing a complex teaching strategy.

There will also be developments in computer-based exploratory environments for children, inspired by AI techniques of programming and knowledge representation, but designed to assist rather than teach. There already exist powerful simulation systems, such as ARK (the Alternative Reality Kit), which allow children to manipulate directly simulations of physical objects. They can create bouncing balls or orbiting planets, and perform experiments on them, 'playing God' by slowing down time or altering physcial laws.

Artificial intelligence and cognitive science are also indirectly affecting education, by providing a language and formalism with which to talk about learning. The notion of 'bugs', for example, described in the section on Logo, has already influenced the teaching of mathematics. Trainee teachers are being helped to identify the causes of bugs in children's arithmetic, such as rules from one process being carried over into another (addition 'carry' rules being used in subtraction, for example), and a diagnostic test for subtraction, widely used in United States schools, was developed by artificial intelligence researchers.

2.4   Military

Advances in the technology of warfare have made dramatic increases in the destructive power of weapons, their complexity and diversity and the speed of their delivery. Today's generals face a much more difficult task than their predecessors in interpreting the battlefield situation and deciding what to do about it. That situation is not only more complex and hazardous, but the decisions must be taken much more quickly and the consequences of an erroneous decision are much more serious. Many military planners have seen the solution to these problems in the greater use of computer technology, and in particular of knowledge-based systems.

KBS is seen as a potential solution to three hard military problems: how to interpret the multiplicity of signals from battlefield sensors; how to operate in hazardous environments; and how to take quick decisions in a rapidly changing battlefield. There are, thus, three major areas of military use of KBSs: signal processing, autonomous vehicles, and battle management systems. For instance, in the United States, a major area of funding for military KBSs is the Strategic Defense Initiative (SDI, or 'Star Wars'). The Strategic Computing Initiative is less well-known than the SDI, but its implications are just as important. It was launched in 1983 by

the US Government as a 5-year project, costing $600 million, to carry out
research and development on three specific military AI systems:

- an Automated Co-pilot (a system that assists a human
  fighter/bomber pilot in tracking targets, etc.);

- an Autonomous Tracked Vehicle (i.e., a robot tank);

- a Battlefield Management system.

2.4.1   Signal processing

In making a successful battle plan, good intelligence is vital. Since situations change so rapidly in modern warfare and there is so much information to collect and collate, automatic intelligence gathering is indispensable.
Raw information comes from battlefield sensors, e.g. radar, sonar, radio,
etc., and must then be interpreted. At the lowest levels this interpretation
consists of grouping the various signals, e.g. deciding that the blips coming
from successive cycles of a radar receiver, or from two separate radar re-
ceivers, originate from the same object. Information of very different kinds
must then be collated, for instance objects identified from radar signals
must be equated with or distinguished from objects identified from their
radio transmissions, visual sightings, journey plans, etc. At higher levels
such objects must be classified as ship, plane, rock, etc., and identified as
friend, foe or neutral. Such signal processing and interpretation systems
form the intelligence gathering part of the battlefield management systems
described below.

   Collating information from very diverse sources of information and dealing with situations that were not explicitly envisaged by the system designers both present very challenging technical problems. The flexibility of KBS
programming techniques offers a potential solution to both problems. KBS
techniques have, therefore, been widely used for the later stages of data
interpretation.

   The best known example of such a KBS system is the HASP passive
sonar interpreter. This system was intended to summarise and monitor
movements of vessels in the Atlantic Ocean using primarily passive sonar
contacts but incorporating sightings, journey plans, and other intelligence
information. Since the time scales involved in ocean warfare are large in
comparison with land or air battles, the system had plenty of time to assess
and interpret the situations. A major breakthrough in data interpretation
techniques would be needed to provide a system with the performance of
HASP for a rapidly altering battlefield.

The diversity of types of information that must be rapidly collated and interpreted and the open-endedness of the battlefield situation, present KBS techniques with a major technical challenge. It is not clear that it is a challenge they can meet.

2.4.2 Autonomous vehicles

Autonomous vehicles could replace manned vehicles in the battlefield. They could continue to operate in environments that were too polluted (e.g. by chemicals or radiation) for humans. Because they are not required to contain and protect human operators they can be smaller and lighter than their manned counterparts. Their use would reduce human casualties (at least, those on the side using them). By making them autonomous, rather than remote controlled, there is no danger of the enemy jamming the control signals from base to vehicle.

Work on autonomous vehicles has proliferated since the Strategic Computing Initiative began, and several groups in the USA and West Germany are working on vehicles capable of navigating unaided over unstructured terrain to achieve a mission goal and return. A detailed newspaper article on the Strategic Computing Initiative[5] has this colourful, and terrifying, description of autonomous vehicles:

> For the army, the defense agency [DARPA] *promises* [our emphasis] an autonomous land vehicle that ... will zip around at 50 kilometres per hour all by itself planning its journey, avoiding obstacles and shooting at the enemy ... DARPA has a few other critters on the drawing board [such as] a security-guard robot dubbed The Prowler.

The major technical problem in these projects is that of system integration. The vehicles typically carry a variety of sensors (laser rangers, colour television cameras, inertial navigation etc.) and have to integrate what their sensors tell them with maps of the terrain, strategic and tactical mission plans, and so on. Collation and interpretation of such diverse information is essentially the signal processing problem described above. Present-day systems are all in the (very) experimental stage and can only go at a few tens of centimetres per second on autonomous drive along metalled roads. Battlefield terrains are usually complex, varied and changing. Major technical advances are required before autonomous vehicles can be built that are capable of coping with these terrains at speeds competitive with their manned rivals. Another aspect of this problem is the design of onboard, autonomous, battle management systems capable of analysing the military situation and deciding what action to take. The problems here are a small scale version of those discussed in the next section.

2.4.3 Battlefield management systems

The role of a battlefield management system is to handle the vast amount of information that arises in modern warfare. Also known as a Communication, Command, Control and Intelligence system (C^3I), it interprets the output of various signal processing programs, collates these with intelligence from other military sources, and presents the results together with the response options in an intelligible form to the responsible military commander. When this commander has chosen a response the C^3I program carries it out by directing the appropriate weapons systems.

Although a human military commander is, in theory, in the 'decision making loop', in practice, the speed and complexity of military decision making often reduces his role to that of confirming the best suggestion of the C^3I system. It is, thus, necessary to program the computer with predefined battle sequences. Unfortunately, battles are open-ended and unpredictable. In fact, one objective of a combatant is to catch his adversary off guard by creating unexpected situations and departing from the normal battlefield sequences. To cope with such unforeseen circumstances a C^3I system would have to be *general-purpose*. It might have to reason about the nature, purpose, and destination of previously unknown objects. It might have to reason about the intentions of an adversary, taking into account the general political situation. It might have to use analogy to cope with an unforeseen situation by adapting an existing plan. Thus battle management, like signal processing and autonomous vehicles, is not the sort of restricted 'toy' domain that lends itself to expert systems technology.

This is not to say that a rule-based C^3I System could not be built; it could, but it would be a special-purpose system not able to cope with unpredicted circumstances. One could regard the problem of detecting and reacting to missile attack as one of fault diagnosis and correction, and adapt an existing expert system shell to do the task. But the resulting system would be no more reliable than a conventional non-KBS C^3I System — in fact less so, since the behaviour of an expert system is inherently less predictable than that of a conventional program. Because of the complexity of the system and the impossibility of testing it in a realistic setting, we can never know whether a battlefield KBS will work properly until it is too late.

2.5 Health

The area of health care offers considerable scope for the use of KBSs to spread expertise more widely, both in the First and the Third World. There are potential benefits in, for example, health education and preventive medicine, medical ethics, the practice of medicine and psychiatry, and the care of the terminally ill.

2.5.1 Medical diagnosis

A great deal of the research work in KBSs has been in the area of medical diagnosis, and many of these projects have had a profound influence on the development of KBSs in general. This must be due, to some extent, to the fact that medicine provides a domain which is rich in problems of different types which provide suitable areas for AI research. However, it is less clear what AI has contributed to medicine, in that there are very few medical expert systems in routine clinical use. There are a number of reasons for this, some of which we will consider later.

One of the earliest expert systems was MYCIN,[6] a system which can diagnose some types of bacterial infection. MYCIN gathered information about the patient by means of a dialogue with the human user, who was assumed to be a trained physician. It asked first for background information, such as the patient's age and sex, and then details of laboratory tests on blood samples. When the responsible organism(s) could be identified, it produced a list of possible therapies and, after a further dialogue, chose the most appropriate one for the patient. Its knowledge was coded primarily as IF/THEN rules, of a form similar to those shown in Chapter 1.

MYCIN was highly influential in that it showed how a rule-based system could achieve impressive performance levels. It was subsequently demonstrated that the domain-specific knowledge (i.e. bacteriology) could be separated from the 'inference engine' (which conducted the dialogue with the user and which could come to new conclusions). MYCIN without the domain knowledge was known as EMYCIN,[7] and was the first of the expert system 'shells'.

An attempt was made to use the MYCIN knowledge base for teaching students how to go about identifying organisms, but the resulting program (GUIDON)[8] was not really successful; on examining MYCIN it was discovered that important knowledge lay hidden, implicit in the structure and orderings of its rules of inference. In order to explain to students what was going on, GUIDON needed to have this knowledge explicitly available. A number of these issues were tackled in the NEOMYCIN[9] program, which, with its derivatives, is still the focus of current research.

MYCIN operated in a domain where the amount of data which could be provided by the user was limited. It could therefore act fairly conservatively, and collect any evidence which might be useful. In contrast, the INTERNIST [10] program dealt with the whole of internal medicine, where the number of signs, symptoms, tests, etc. which could be asked about runs into thousands. The developers of INTERNIST had to look very carefully at the ways in which clinicians went about deciding what hypothetical diagnoses to consider and, as a result, what questions to ask.

Another important program was CASNET, [11] which assisted doctors in diagnosing glaucoma. CASNET is interesting in that it attempted to reconstruct the history of the disease, and to predict its future development. It too gave rise to a shell called EXPERT, which was used to construct the CLINISCAN system. This is an expert system built into a laboratory instrument to carry out serum protein electrophoresis analysis. It is probably significant that most of the data required is available within the instrument, and the user can obtain the advice with very little extra effort; this should be contrasted with some of the early systems which required a considerable amount of extra typing. Indeed, we can speculate that medical expert systems will not develop their full potential until computerised medical records are widely available — and that raises problems, of quite another kind, in the field of privacy.

Two other systems which were developed by the MYCIN group were PUFF [12] and ONCOCIN. [13] Like CLINISCAN, both systems have large amounts of data already available. PUFF is used routinely in one hospital to analyse pulmonary function tests; the clinician responsible for the unit routinely signs 80 per cent of PUFF's reports without further comment. ONCOCIN, on the other hand, is still under development. Its objective is to advise on adherence to cancer therapy protocols, and is integrated into a data collection system which was specifically designed to replace the existing paper records. Despite many person-years of effort, ONCOCIN is still not widely used; this is in part due to the problem of funding the transfer from a research prototype to a fully supported operational system.

Most of the systems described above operate in 'consultation' mode — i.e. the user comes to the system with a problem, and, after some interaction, the system provides a 'solution'. A very different approach was pioneered in the ATTENDING [14] system, whose domain was that of the management of anaesthesia. The user of ATTENDING not only describes the problem but also provides his/her own solution. The system then 'critiques' the solution by comparing it with its preferred plan. The basic idea behind ATTENDING was that expert systems should not act as all-knowing oracles which deliver a judgement from a great height, but rather as intelligent assistants who can be used as a 'sounding board' for ideas.

There is a great deal more research to be done in this area, but it seems clear that, in medicine at least, a critiquing system is likely to be more acceptable.

We have already mentioned some of the factors which have militated against the widespread use of expert systems in medicine. One other factor is likely to be the limited scope of the knowledge which they encode. Most of this is in the form of surface associations between manifestations and diseases. However, in certain medical specialties, there is considerable understanding of the underlying mechanisms, in terms of anatomical structures and physiological mechanisms. Finding ways of representing and reasoning with such knowledge is not simple, and is likely to take some considerable time to achieve.

2.5.2   Primary health care

In Britain, as in many industrialised countries, there has been a greater emphasis in recent years on preventive medicine and the provision of health education. GPs, health visitors and district nurses have traditionally been responsible for this type of health care, and with a rise in the number of elderly people and a policy of caring for the sick and disabled within the community their role has become more important. The district nurse, health visitor, midwife, and other health specialists now work together as a team, alongside the general practitioner, with collective responsibility for the health of a local area. Increasingly, they are expected to offer advice on preventive medicine, diet, family planning, pediatric health care, and psychiatric problems and this demands both specialist knowledge, for example on nutrition, and wide background information on referral agencies and back-up services.

Some of this need for knowledge and information could be satisfied by conventional information sources, from a card index to a computer database, but a KBS could provide useful support by allowing the user to follow health trends within a family or the wider community and to produce a plan of action for the health team. A health visitor equipped with a KBS could, for example, follow the course of a whooping cough epidemic and examine possible strategies for advice on immunisation.

District nurses are trained in the detection of pain, and so offer advice to the family doctor on changes in drug dosage levels. What they lack is the specialised knowledge of drugs that would enable them to make detailed recommendations. A KBS could not only provide such information, but it could also show the possible effects, over time, of different drugs and could plan a course of medication.

The flow of knowledge is not just in one direction. Community health workers gain practical experience that should be more widely disseminated. One health care program intended to meet this sort of need is a KBS being developed by the Imperial Cancer Research Fund which will guide the prescription of pain-killing drugs to terminally ill patients. The relevant expertise comes from the hospice movement, whose specialist knowledge covers not only what sort of personal care can protect the dignity and serenity of the dying but also which drugs may be necessary to make this dignity possible. These drug treatments have been discovered in hospices, not in hospitals, and are not yet part of the medical curriculum. Consequently, a KBS in this area could be very useful if it were made available to hospital consultants and GPs.

Another example of a KBS dealing with aspects of health care that are often skimped in the training of medical and nursing staff is a program — already being distributed by the B.M.A. — which embodies the rules of law and generally accepted medical ethics relating to 'consent'. That is, it clarifies the various issues involved in deciding whether a particular patient, or some suggested proxy, is competent to give valid consent to medical treatment. The B.M.A. hopes that it — and a similar KBS (under development) on ethical problems of medical confidentiality — will be used as a teaching aid for medical, paramedical, and nursing students, and as an occasional refresher course for established professionals. Further, they recommend it to medical practitioners as an aid to their decision-making in particularly difficult or complex cases. [15]

2.5.3   Psychotherapy

Some health-care personnel are exploring the use of KBSs in the initial stages of psychotherapy — specifically, to help identify the patient's areas of concern as quickly as possible. In a large psychiatric hospital in California, for instance, patients may be offered the opportunity to be 'interviewed' by a computer. The program is not a rule-based KBS, but a natural-language processor which can recognise (and generate) some words or phrases dealing with psychologically sensitive issues, such as emotions (anxiety, depression, guilt, happiness) or family relations.

The patient has a 'conversation' with the program, which from time to time asks questions about topics likely to be of therapeutic interest. Psychiatric staff are present throughout this 'interview', though they will look over the patient's shoulder (or help with the typing) only if invited to do so; other patients may be in the room also, interacting with the same program on other keyboards. The interview is printed out after it has finished, and is later discussed in detail by the psychiatrist and the patient together.

Patients commonly introduce psychologically sensitive topics at an earlier stage than if the interviews had been conducted with a human psychiatrist. Some patients say they find it less embarrassing to raise anxiety-ridden themes with the program than with the doctor, for the program (unlike the human) cannot be shocked and will not pass any moral judgements. Moreover, the use of the print-out sometimes seems to help the patient to discuss things in a more objective, or clearer, way than usual.

But a psychiatric interview differs from a dialogue to diagnose a physical problem in that we cannot assume the patient is a reliable respondant. If a person says 'my stomach hurts' you can assume that there is some physical symptom to be further investigated, and that the investigations will give more objective evidence to the claim. If a person says 'my mother hates me' in the context of a psychiatric interview then it is not appropriate simply to assign a confidence factor to the statement and then look for supporting evidence. A person's mental construct of themselves is not a static object to be analysed or modelled, but a consequence of an active participation in a dialogue. Alter that dialogue, by substituting a computer for a person, and the construct alters. In psychotherapy, we are interested primarily in how people relate to people, not to machines.

There is a widespread view that it is immoral to use a computer for a function such as psychotherapy that relies essentially on respect, understanding, empathy or love between humans. A program might be able to construct convincing responses, which lead a patient to divulge guarded information or feelings, but the patient's trust would be entirely misplaced. The program should not offer psychiatric remedies; at most it should act as a conduit to a human psychotherapist.

It would appear that the computer can be used constructively *as part* of a psychiatric interview, but only if it assists the immediate task, such as eliciting responses from the patient, and also does not exacerbate underlying problems such as alienation or fear of technology.

2.5.4   Conclusions

Since the aim of using current medical knowledge to cure disease or prolong life is one which most people (except Christian Scientists) share, it is not surprising that a number of medical expert systems have been and are being produced to aid the doctor's diagnosis and treatment. But the doctor's decisions about what to do may be subject to financial pressures (on the patient or the Health Service), and should take into account the relative acceptability to the patient of the quality of life provided by the treatment. Financial and personal pressures of this kind are not represented in current medical KBSs, which can advise only on purely technical matters. Possibly, some future KBSs might be used to help hard-pressed medical personnel to make choices involving scarce resources — of money, time, or donated organs — but the ethical issues involved should be represented at least as well as they are in the consent-KBS mentioned above.

References

1. Gemignani, M. (1987). The regulation of software. In *ABACUS*, Vol. 5, No. 1, pp. 57–9.

2. Kay, H., Dodd, B., and Simer, M. (1968). *Teaching machines and programmed instruction.* Pelican, London.

3. Sharples, M. and O'Malley, C.E. (1988). A framework for the design of a writer's assistant. In *Artificial intelligence and human learning: intelligent computer aided instruction*, (ed. J. Self). Chapman Hall, London.

4. Self, J. (1985). *Microcomputers in education: a critical appraisal of educational software.* Harvester Press, Brighton.

5. *San Jose Mercury News West Magazine*, November 23, 1986.

6. Buchanan, B.G. and Shortliffe, E.H. (1984). *Rule-based expert systems: the MYCIN experiments of the Stanford heuristic programming project.* Addison-Wesley, Reading, Mass.

7. Van Melle, W. (1979). A domain-independent production-rule system for consultation programs. In *IJCAI-79*, vol. 1, pp. 923–5.

8. Clancey, W.J. (1982). Tutoring rules for guiding a case method dialogue. In *Intelligent tutoring systems*, (eds. D.H. Sleeman and J.S. Brown). Academic Press, London.

9. Clancey, W.J. and Letsinger, R. (1981). NEOMYCIN: reconfiguring a rule-based expert system for applications to teaching. In *IJCAI-81*, vol. 2, pp. 829–36.

10. Miller, R.A., Pople, H.E. and Myers, J.D. (1982). INTERNIST-1, An experimental computer-based diagnostic consultant for general internal medicine. In *New England Journal of Medicine*, vol. 307, no. 8, pp. 468–76.

11. Kulikowski, C.A. and Weiss, S.M. (1982). Representation of expert knowledge for consultation: the CASNET and EXPERT projects. In *Artificial intelligence in medicine*, (ed. P. Szolovits). pp. 21–56, Westview Press, Boulder, Colarado.

12. Aikins, J.S., Kunz, J.C., Shortliffe, E.H. and Fallat, R.J. (1982). PUFF: an expert system for interpretation of pulmonary function data. In *Computers and Biomedical Research*, vol. 16, no. 3, pp. 199–208.

13. Shortliffe, E.H., Scott, A.C., Bischoff, M.B., Campbell, A.B., Van Melle, W., and Jacobs, C.D. (1981). ONCOCIN: an expert system for oncology protocol management. In *IJCAI-81*, pp. 876–81.

14. Miller, P.L. (1983). Medical plan-analysis: the ATTENDING system. In *IJCAI-83*, pp. 239–41.

15. Sieghart, P. and Dawson, J. (1987). Computer-aided medical ethics. In *Journal of Medical Ethics*, **13**, pp. 185–8.

3 The benefits and dangers of KBSs

3.1 Attitudes towards KBSs

Artificial Intelligence has been a powerful theme in literature, from the automaton encountered by Jason and the Argonauts, through Frankenstein's monster, to the robots of Isaac Asimov. The image of the 'thinking machine' still evokes both fascination and revulsion, and these attitudes will have an important influence on the future benefits and dangers of KBSs. Even those involved in research and development are not above such passions; many of them entered the field enthused with the idea of designing a mind, and the Lighthill Report [1] even suggested that robotics researchers suffer from maternal deprivation and build robots in the image of their mothers! (In 1972 Sir James Lighthill was asked by the Science Research Council to prepare a report on the current state and prospects for Artificial Intelligence. His extremely critical views all but halted AI research in Britain.)

The survey of AI researchers [2] mentioned in Chapter 1 found three main clusters of attitudes: the Believers, the Social Critics and the Disbelievers. The Believers consists of those:

> who are strongly optimistic about the results of AI research and see profound (mostly positive, but in any case inevitable) social changes as a result. These changes include relieving humans from nearly all routine or menial tasks and perhaps many of the more complicated tasks as well as: helping make more rational decisions in government or the judicial system; improving and customising education; and helping to empower citizens and democratize society through more powerful computers accessible to the layman.

The Social Critics hold the view that:

> although AI may eventually succeed technically at much of what
> it attempts, many of the social implications will either be nega-
> tive or will simply reinforce existing power structures in society.

The Disbelievers argue that:

> many of the goals of AI are not possible at all, or not possible
> in the way they are framed by researchers. H.L. Dreyfus [3] is the
> most visible of these writers. Dreyfus argues that fundamental
> progress in AI has been stymied by what he considers four er-
> roneous assumptions in their approach to a thinking machine:
> that a digital computer resembles the brain in the way that it
> handles information; that the brain processes information as a
> computer does, at some level; that human knowledge and be-
> haviour is formalizable; and that knowledge can be meaningful
> in discrete chunks.

These groupings are caricatures, and the majority of AI practitioners
would probably hold some combination of these attitudes. Certainly it is
quite possible to be both a Believer and a Social Critic. Nevertheless, the
groupings provide a useful framework for a discussion of the benefits and
dangers of KBS.

3.2 The believers

3.2.1 Cheap and ready knowledge

Much of the optimism about KBSs is not generated by their proven capability, but by their potential. The belief is that considerable benefits could be
brought to industry, government, medicine, education, voluntary groups,
and ordinary people in their homes, by the cheap and easy availability of
knowledge on a wide range of topics.

Many problems arise through ignorance. We suffer, for example, because
we do not know how to cure our sicknesses, or how to become richer, or
where to go for help. The information we require may well be available,
but hidden in legal or medical tomes, or in the heads of professionals.

A Third World farmer may have plants ravaged by disease, but not
know what the disease is, or how to cure it. This knowledge may not be
available in the farmer's village at all, nor elsewhere in the country, at an
affordable price. An expert system for diagnosing plant diseases could assist
in identifying the disease, and curing it, for a reasonable cost.

An unemployed person may not have enough money to support himself.
Although that person may be entitled to more benefits, he or she may

not be claiming them because of a lack of independent advice. An expert system on DHSS regulations could advise on rights and how to go about claiming them.

A small company designing a new gadget might not be able to afford the investment in time and effort required to make the gadget as efficient, and hence as competitive, as possible. An expert system might assist in the design by, for example, calling up mathematical modelling programs, or giving expert advice on the design choices.

3.2.2 Sailing the sea of information

It is not more information that people need — we are swimming in an ever-rising sea of information, from news bulletins, surveys, quiz shows and mail-shots — but knowledge directly relevant to a problem: knowledge about which set of actions is appropriate to the circumstances, about how to perform a task, and about the consequences of each possible decision. At present specialist knowledge is very unevenly distributed, concentrated in universities, teaching hospitals, company head offices, government departments, and specialist advice centres. One way to ensure its spread is to provide non-specialists, such as school teachers, general practitioners and sales representatives with a source of knowledge applicable to their immediate needs, in the form of an expert system supplemented by a referral service to the human expert for the particularly difficult cases. There is an obvious danger that the non-specialist may not be able to spot a problem requiring the intervention of a human expert, but the alternative to the limited expertise of a non-specialist plus expert system may be no expertise at all.

The 'sea of information' brings its own problems. Only a tiny fraction of what we might see, hear or read is directly useful, or even interesting, yet finding those nuggets of information could involve hours of tedious search. The Prestel information system, operated by British Telecom, contains some 100 000 'screens' of potentially useful information, but these are spread amongst hundreds of 'information providers': independent companies who promote their own data and ignore those of their competitors. To find, say, the availability of skiing holidays in Italy involves searching through the databases of ten or more travel firms, all using different methods of classification and different formats for displaying the information on the screen. Expert systems offer some hope of managing the spread of information by turning it into usable knowledge in the form of digests, decision options, and automated monitoring systems. Programs already exist that automatically monitor and index the news stories provided by press agencies, providing stories to order on a given topic; there are also ones that collect data on stocks and shares, giving a warning of any large rise or fall in selected stocks.

3.2.3   The human window

Donald Michie has warned of situations where a sudden 'flood of information' can completely overwhelm a human monitor. During the emergency at the nuclear power plant at Three Mile Island over 100 warning lights were flashing simultaneously, the operators wrongly interpreted the warnings and, instead of drenching the reactor core with water, switched off the emergency cooling system (which the computer had correctly activated), with disastrous results. But the computer program at Three Mile Island was not an expert system and merely set off the cooling system, with no explanation of its actions. Michie argues that complex systems like power plants, or chemical installations, or aeroplanes, need a 'human window'[4] which, in an emergency, can give the human operator an intelligible précis of the situation and in particular *why* the program is doing what it is doing. Expert systems have the ability to interpret large amounts of data, turning it into usable knowledge to guide human action.

3.2.4   The general factotum

From all this it may seem that expert systems are only for use in an emergency, but a portable and knowledgeable assistant could be the modern equivalent of a 'general factotum', replacing do-it-yourself books, car maintenance manuals and self-study guides by offering knowledge in the context where it is needed. Devices such as cookers or car engines may, in future, contain their own source of expertise. Already there exist packages of electronics capable of directly monitoring car engines, and reporting faults. Coupled to an expert system one of these could not only provide a warning of an engine fault, but also offer advice on a remedy. Each domestic appliance may in the future be a little more intelligent, with the central heating system automatically minimising heat loss, the oven recommending the best cooking method for a particular dish, and the front door telephoning to announce that it has inadvertently been left open.

If the thought of accepting advice from an oven or conversing with a door appals, then the Believers offer the consolation that each wave of innovation — the train, the motor car, electronics — has, in the short term, brought with it dissonance and disruption of familiar patterns of work and leisure, but has eventually been assimilated and accepted.

In summary, the Believers suggest that everyone will benefit in some form from the widespread use of KBSs, but the people with most to gain

are those who are currently denied access to specialist knowledge through
its cost or their lack of power, those who rely on expert knowledge as part
of their work, and those who are attempting to cope with complex fast-
changing systems.

3.3   The social critics

3.3.1   KBS literacy

As a result of over 3 000 years of literacy most people in the industrialised
countries, and increasingly many in the Third World, are now able to read,
and to detect some of the lies, assumptions and fallacies buried in text.
Even so a good writer can bend the mind of his or her reader, with a
misleading analogy, a subtle distortion, or some false logic. Commercial
expert systems have been with us for a mere five years, and there is no
equivalent base of literacy. Users have not learned the skills of 'reading' the
rules of expert systems, and uncovering false assumptions or detecting gaps
in the knowledge. Such literacy will evolve, but not in the near future,
since the structure, methods of representing knowledge, and methods of
reasoning used by KBSs are still changing.

The result is that a user of a KBS must take its conclusions largely on
trust; as we mentioned in Chapter 1, many expert systems have rudimen-
tary rule justification and explanation facilities, but these do little more
than list the rules used to form a conclusion. They cannot comment on
whether each rule is a correct representation of the real world, nor whether
the reasoning itself is complete and reliable. Consider the example rule
from the Wine Advisor:

```
Rule colour1
IF the colour of the wine is unknown
AND the main dish is meat
AND the main dish does not contain veal
THEN
    remove the fact that the colour is unknown
    add the fact that the colour of the wine should be red,
    giving it a certainty of 0.9
```

If the 'not' had been missing from the third condition, ie:

```
AND the main dish contains veal
```

then the rule would still operate, only this time it would fire when the main dish *contained* veal. That mistake might be hard to detect (since red wine *is* appropriate for drinking with veal), but now the certainty factor is all wrong: another rule should be fired when the dish is veal. If the 'colour1' rule had been missed out altogether then the system would still run using other 'default' rules (one of the proclaimed advantages of expert systems over other computer programs is that they can still operate if some of the data or knowledge is missing), but the conclusions would be different, in some cases seriously wrong. And this is only the simplest of expert systems. The consequences of missing or faulty rules in a large expert system, perhaps monitoring a nuclear power plant, might be not a poor choice of wine but a meltdown.

3.3.2 Closed windows

The benefits, mentioned in the previous section, of providing a 'human window' in the form of a KBS that can monitor and offer advice on a complex process, rest on the assumption that the KBS is itself understandable and reliable. If it is not (and the first sign that it is faulty may come in the very emergency it was supposed to assist) then the human user may not be able quickly to call up the raw data, or may have lost the skill in interpreting them; he may not even be able to tell whether the fault has occurred in the underlying system or in the KBS. Thus, the term 'window' is highly misleading; a monitoring KBS is no sheet of glass, but a complex system, interpreting another complex system to the human user. Adding an expert system to a nuclear plant may give the appearance of simple efficiency, but so would removing 90 per cent of the gauges and warning lights.

3.3.3 A little knowledge

Making KBSs more like human advisors — for example, by enabling them to converse in something like natural language, or to employ common-sense reasoning — runs the risk that they will be taken as more like human experts than they actually are. As a result, people may put undue trust in their pronouncements.

For instance, a future KBS could appear to have a fairly subtle command of natural language within the limited interchanges for which it was designed. Many users might therefore assume that it has a complete command of that language, at least in the subject area under discussion. Some people might even believe it to have a rich command of language in other areas too. These false assumptions would very likely lead to its judgements being given much more credit than they are worth.

Suppose, for example, that the computer uses a familiar English word such as 'possible'. The human user knows that this word is similar in meaning to a number of others, such as 'probable', 'credible', 'hypothetical', 'arguable', 'natural', 'foreseeable', 'likely' and so on. But we know too that 'possible' is not precisely equivalent to any of these, for each of these words has a subtly different shade of meaning. If a native English speaker uses a particular word, we can assume that he or she has chosen it *in preference* to any of the alternatives. This is why someone who says that something is 'possible' is assumed to be denying that it is probable. For a KBS the word 'possible' may be the only one which covers all the above alternatives, with a range from 'likely' to 'arguable'. The more superficially 'human-like' the KBS, the more people may be misled into thinking that it possesses a full range of human attributes such as common-sense reasoning and language understanding.

3.3.4   Autonomous systems and the technical fix

An expert system designed to interpret a large quantity of information would normally leave the human user in charge, able to make the final decision on whether or not to act on the system's recommendations. But there are circumstances, for example interpreting fluctuations in exchange rates or share prices, analysing data from meteorological stations, monitoring complex industrial processes, or analysing signals from military early warning systems, in which the information is changing so rapidly, and decisions need to be made so frequently, that a human cannot keep up. Even if a human were, in theory, given the final say, his role would only be to confirm the best guess of the system or to make a random choice between options. There would not be time for a more informed decision. For these situations, KBSs are being designed that not only provide recommendations but also activate machinery to carry them out: automatically buying and selling shares, creating weather maps, controlling industrial processes, or launching a retaliatory military strike. It is autonomous systems for military use that raise the most immediate social issues, since the consequences of a failure are so great. A failure in the midst of battle could significantly lower the chances of winning. A failure in peace time might cause an unintended war.

Unfortunately, the likelihood of failure is high. All computer programs contain bugs; the bigger the program the more bugs it contains. KBS programs are no exception. Bugs can arise either because the original specification of the program did not anticipate all the situations that the program will be faced with or because the specification has not been fully met by the implementors of the program.

Programs are made safe for normal use by an extensive process of testing and correction. For programs of any complexity it is infeasible to test them in all possible situations. This is doubly so when the situation they face is open-ended, so that some of the possible situations they may face are not known to the system's designers or testers. Military programmers face an adversary who is constantly trying to think up such unexpected situations to trick the program. In such cases the program cannot be properly specified and, thus, will contain bugs however well it is implemented. In addition, some military programs, e.g. those that react to enemy missile attacks and launch retaliatory strikes, cannot be tested in a real situation, but only in simulations that may be flawed by false assumptions and limited imagination. Bugs of both specification and implementation will remain undetected and uncorrected.

An official Department of Defense document acknowledges problems with such systems:

> Commanders remain particularly concerned about the role that autonomous systems would play during the transition from peace to hostilities when rules of engagement may be altered quickly. An extremely stressing example of such a case is the projected defense against strategic nuclear missiles [presumably a reference to the Strategic Defense Initiative], where systems must react so rapidly that almost complete reliance will have to be placed on automated systems. At the same time, the complexity and unpredictability of factors affecting decisions will be very great. [5]

Despite such official worries, there remains the underlying assumption that automated military decision-making systems are inevitable. This is a supreme example of a 'technical fix': as a result of the development of short and medium range missiles, the 'response time' (the time between detecting a missile attack, and the missiles exploding on target) can be as little as three minutes. In that time a decision must be made as to whether to launch a counter-attack, before missile silos are destroyed. The 'technical fix' is designed to cut out the human decision-maker, and to adopt a 'launch on warning' strategy, so that as soon as the computer detects incoming missiles it automatically commands a counter-attack.

It is not clear whether the U.S.A. has now adopted a strategy of launch on warning. Dr Clifford Johnson, a Stanford University computer scientist, claims that it has. According to Johnson, the United States Air Force has stated that a 'launch on warning capability' is essential because to risk the loss of land-based missiles in a first strike by Soviet forces would undermine deterrence. Dr Johnson has filed a court suit against the Secretary of

Defense, claiming that this policy breaches the U.S. Constitution, in that it subverts a number of provisions regarding legally authorised sub-delegation of powers: the Congress cannot delegate the power to declare war to the President, and the President cannot delegate the power of the command of the armed forces to a computer.

Faced with the complexities and speed of modern warfare, military planners are being forced down the road of increased automation. They have been seduced by the apparent flexibility and 'intelligence' of KBS systems and think they have found a 'technical fix' for their problems. Unfortunately, current KBS techniques are not up to the job. Their successes so far have been in circumscribed domains very different from the diverse, open-ended and adversarial military ones. If there is no 'technical fix' then we must look for a 'political fix' which defuses the dangerous military confrontation.

3.3.5   The emotional KBS

So far, the criticisms of KBSs might be construed as technical objections, to be overcome when the next, or the next but one, generation of KBSs arrives: systems that can learn from experience, converse in fluent English, and be praised for their quick thinking and common sense. Whether or not such machines will ever be built is open to question, but assuming they can be, problems may still remain.

People have become accustomed to computers producing the truth. One may doubt the accuracy of a bank statement, but not the arithmetic associated with it. Computers are often accused of making errors, but the public is being trained to blame the data, or the programmer who has specified the wrong set of instructions, not the computer itself which slavishly follows the commands.

But now all this carefully nurtured public acceptance must be thrown out; KBSs are *designed* to be inexact. Their output must be judged by the criterion of *adequacy*, not correctness. Indeed, Sloman [6] argues that an inevitable consequence of designing larger and more general-purpose systems is that they will behave as if governed by human characteristics, such as uncertainty, belief, and anxiety:

> It can also be argued that there is no way to build a super-intelligent robot which also copes with a complex set of different sorts of motives, in a partly unpredictable world, without giving that robot mechanisms which are capable of producing emotional states, as a result of performing the cognitive tasks for which they are required. That is, the possibility of having

emotions may be a by-product of being able to cope with a complex and unpredictable world in an intelligent way. (This does not mean that every intelligent robot will necessarily be emotional, only that it will have the ability — and abilities are not always exercised).

Emotions are not necessarily harmful or unproductive, but they do not square with the image of computers as efficient and accurate. A KBS that is designed to be uncertain or emotional may not win public acceptance, particularly if its decisions may affect people's welfare or, in the case of military KBSs, threaten their lives.

Alternatively, people may come to believe and trust the pronouncements of KBSs. The uncertainties involved in reaching a decision can be easily hidden, with the system reporting only its final conclusion. This is less likely in a KBS which acts as a decision support system; even with present-day systems the human user can demand justifications and alternative suggestions. But once KBSs are designed to take action based on a decision, they may appear to perform *as if* driven by the truth. Imagine a system in some hospital of the future which, given the diagnosis of a patient's condition (perhaps provided by another expert system), will not only prescribe drugs but also administer them. All kinds of difficult decisions must be taken — whether to use a drug with general or specific action, what dosage level to give, perhaps even what drugs can be afforded given the current state of the hospital's finances — but all that the patient receives is a shot in the arm.

3.3.6   Hidden rules

The main beneficiaries of KBSs, at least in the short term, are those who already use computers to process information, who require more effective ways of extracting useful knowledge from the data, and who can pay the large cost needed for developing a KBS to do the job. (According to a recent report [7] the production of a system with about 300 rules takes one to two years and can cost up to $1 million.) Detailed knowledge about large numbers of people can now be extracted from computer databases containing, for example, credit information, police incident reports or social security documents. One of the most widely articulated fears about computers relates to the dangers which they pose to personal privacy. Those dangers are indeed very real, and they have now been well documented. (For a general overview of this subject, see Sieghart, P. (1976). *Privacy and computers.* Latimer, London.)

As a result, many countries have enacted data protection legislation.

Typically, such legislation — as in the case of the UK Data Protection Act — requires people ('data users') who use computers to process personal data about identifiable individuals ('data subjects') to disclose on a public register what kinds of personal data they hold, for what purposes they hold them, and to whom they disclose them. They are also required to comply with a set of 'data protection principles'; to tell the data subject, on request, what personal data they hold about him; and to correct those data if he shows that they are wrong or misleading.

Data protection legislation draws no distinction between different kinds of computer system, and any KBS which processes personal data will therefore fall automatically within its scope. That being so, KBSs ought to present no greater threats to personal privacy than any other kinds of system. However, they raise an interesting issue which, so far as we know, has not been debated in the context of data protection, namely whether data subjects should be entitled to know not only what data are held about them for what purposes, and to whom they are disclosed, but also the *rules* by which they are processed. For instance, suppose a credit rating KBS contains a rule of the general form

IF subject address is in postal district Brixton
AND subject age is less than 25
AND subject race is non-caucasian
THEN subject credit rating is poor

Should the data user be bound to disclose such a rule on the data protection register, and tell his subjects about it on request? One could clearly make a strong case that he should — but, at present, data protection legislation does not expressly require him to do so. Without knowledge of the rules by which data is to be processed, a data subject is unable to assess whether a computer system is providing accurate information about him. For instance, it may provide as apparent fact information that is actually derived from accurate data using dubious rules.

Such rules can, of course, be programmed implicitly into any computer system. A blatant example of such practice comes from the report of the Commission for Racial Equality into St George's Hospital Medical School. [8] They found that the computer which the School used to make an initial selection of undergraduate applicants was programmed to discriminate against women and ethnic minority candidates, giving them a less favourable weighting and thus a lower probability of being interviewed. The program had deliberately been written to select the same set of applicants as the human doctor had; *implicit* bias against women and ethnic minorities had already been in operation, and this was emulated in the computer program.

Between 1982 and 1986 all selection for interview was done by the computer program. The bias was not immediately detected because the workings of the program were not freely available and the bias was not explicitly expressed as a rule. However, once bias *was* suspected it was readily verified by changing the race and/or sex of an unsuccessful candidate and running the data through the program again. It would not have been so easy to verify bias in a human selector, and indeed the bias of the human selectors went undetected for many years prior to 1982.

Public access to program rules seems to us to be particularly important for future KBSs in the public sector, and especially those designed to identify data subjects about whom adverse inferences might be drawn. Any legislation would have to balance the public 'right to know' and the desirability of detecting prejudice and bias, against the rights of the program owner when secrecy was legitimate. For instance, the rules used by a company might be regarded as a commercial secret giving it a legitimate advantage over its rivals. A law enforcement agency might argue that knowledge of its rules would enable an offender to evade detection.

There is no universally acceptable solution to this conflict of interest: the simplest response would be to extend the provisions of the Data Protection Act to cover not only data, but also all explicit rules by which data are processed, so that people would be allowed to inspect those rules that applied to them as 'data subjects' (for example in a computer system that assessed individuals for job interviews). The Data Protection Act has provisions for certain classes of sensitive information to be kept secret from the general public.

The issue is most acute in the law enforcement field, ranging from police intelligence systems to those designed to detect social security frauds. It is not too difficult, for example, to envisage a system which would bring together the 'rules of thumb' by which police officers identify various kinds of potential villain. A recent book by a Deputy Assistant Commissioner to the Metropolitan Police [9] spells out some indicators of suspicious characters:

- young people generally, but especially if in cars (and even more so if in groups of cars);

- people in badly-maintained cars, especially if they have a tatty dog-eared licence;

- people of untidy, dirty appearance, especially with dirty shoes;

- people who are unduly nervous, confident, or servile in police presence (unless they are doctors, who are 'naturally' confident);

- people whose appearance is anomalous in some way — e.g. their clothes are not as smart as their car;

- people in unusual family circumstances;

- political radicals and intellectuals, especially if they 'spout extremist babble'.

According to the book, normal unsuspicious people are those outside these categories, especially if they are of smart conventional appearance, and even more so if they smoke a pipe. It is quite easy to imagine these indicators being included as rules in a police KBS. It is also easy to imagine someone who happens to display some (or even all) of the suspect characteristics, and yet is a perfectly law-abiding citizen. Such a person would probably already be the object of suspicion by his local constabulary, but his situation could be made much worse if an 'expert' computer system had 'assessed' him as a criminal with a 'confidence level 0.9'. On the other hand an explicit representation of police rules of assessment may assist in the correction of injustices. The main problem is one of unreliable indicators, not of their embodiment in a machine, although machine rules can make the situation both worse, by giving a false judgement more credibility, and better, by making it easier to detect.

3.3.7   Selling knowledge like Coca Cola

Even if the fears prove groundless, and even if KBSs become highly-regarded assistants and colleagues, there is still one problem. Inevitably the spread of technology will not be uniform. The industrialised and technologically sophisticated nations will be the first to design and sell such tools for the mass market. The remaining countries, particularly in Africa, Asia, and Latin America, will increase their dependency, not only on high technology but also on alien assumptions about the content, structure, and purpose of knowledge. An agricultural KBS, for example, may suggest solutions in the form of proprietary brands of feedstuff, or may be governed by questionable assumptions about the need for artificial fertiliser.

In the view of the Social Critics KBSs may well work, but the consequences of the widespread introduction of such technology, into all aspects of our lives, have not been understood or even properly discussed. Until we can be sure that they will be beneficial, they argue, we must not rely on complex and *necessarily unpredictable* systems to guide and control our already over-complicated lives.

3.4   The disbelievers

The camp of Disbelievers is smaller than the other two, partly because those approached in the survey cited earlier were all involved in research related to AI, and so hardly likely to cut their own throats, but also because no-one has yet offered a formal proof that general machine intelligence is impossible. It has been shown that for any particular logical or mathematical system there is at least one statement that cannot be proved or disproved within the system. This is certainly a theoretical limitation of any particular KBS, but not of machine intelligence in general: the Believers do not claim that a single system can answer every question posed to it, but only that KBSs will one day show mental abilities equal to or surpassing those of human beings.

The most prominent of the Disbelievers, Hubert Dreyfus, does not rest his case on the 'incompleteness' argument, but begins by pointing out that although present-day KBSs may superficially appear to have 'human-like' intelligence, they do not reason in a human-like way. They perform like narrow specialists totally lacking in common sense, and they employ techniques and representations that have no parallel in the human mind:

> As we see it, all AI's problems are versions of one basic problem. Current AI is based on the idea which has been around in philosophy since Descartes, that all understanding consists in forming and using appropriate representations. Given the nature of computers, in AI these have to be formal representations. So common sense understanding has to be understood as some vast body of propositions, beliefs, rules, facts and procedures. AI's failure to come up with the appropriate formal representations is called the common sense knowledge problem. As thus formulated this problem has so far turned out to be insoluble, and we predict it will never be solved. [10]

Hubert Dreyfus and his brother Simon argue that KBSs might perform competently in certain narrow domains, but will never become equal to human experts, since expertise also demands common-sense reasoning, and possibly 'holistic processes quite different from the logical operations computers perform on descriptions'. As examples, he mentions the ability of the human mind to scan and rotate mental images, and to recognise the similarity between whole images:

> Recognising two patterns as similar, which seems to be a direct process for human beings, is for a computer a complicated process of first defining each pattern in terms of objective features

and then determining whether by some objective criterion the set of features defining one pattern match the features defining the other pattern. [10]

But these arguments apply to the current generation of computers, the result of a mere 40 years of research. What may be achieved in the next 40 years is anyone's guess, but already a new range of computing systems is under development, based more directly on the structure and operations of the human brain. Early results suggest that recognising similarities is one task that they are able to do much better than 'orthodox' computers, and in a way that appears to mimic the brain.

Although the Disbelievers are right in stating that AI has not lived up to the more extravagant early claims, or even some of the more sober hopes (for instance that simple expert systems could be easily extended, to cope with far more knowledge and rules), they have no knock-down argument that AI is impossible.

3.5   All in the mind

At least in the short to medium term, the course of development of KBSs is likely to be dictated not by technical limitations, but by public attitudes and commercial acceptance. The public may embrace them, as it has embraced video games and home computers, or people may feel threatened by the challenge to their own skills and expertise.

At present, many people feel that comparing the mind with machines must inevitably be dehumanising. They assume that human beliefs, purposes and freedom — in a word our *subjectivity* — will be rejected as illusions. For after all, what human minds can do can be done also by KBSs which, such people assume, are purely *objective* systems: mechanical reasoners without prejudice or subjectivity. If computers can reason, plan, and choose with no need of subjectivity then (so the argument goes) the subjectivity of human minds is at best an irrelevance to most of our mental life, at worst a sentimental illusion. Much as many people in the nineteenth century felt Darwin's theory of evolution to be dehumanising, so many people today believe that a science of the mind which compares us with computers thereby denies our humanity.

But as we have seen in the previous chapters, KBSs are not 'objective' in this sense: they function as *models* or *representations* of the world, and of possible worlds. As such, their 'judgements' are essentially fallible, and open to challange. It is true that a KBS does not really *understand* what meningitis — or anything else — is, as we do. A programmed computer is not a mind. But it is sufficiently like a mind, in its use of internal representations, for it to be treated as a *non-objective* system. Considered as a piece of electronic machinery, a computer is objectively a part of the world (as the brain is too), but considered as a symbolic system or program, it is a set of subjective representations of the world (as the mind is too).

The analogies between computer systems (of various kinds) and human minds are stressed by psychologists seeking a scientific understanding of the mind. For psychologists the computer has become an important tool in the development of hypotheses about mental states and processes. Although a computer model of the mind does not constitute a proof of some psychological function, it does suggest possible thought-mechanisms that can then be tested by traditional psychological experiments. By building (or trying to build) a program that learns, or understands natural language, or recognises objects in a scene, psychologists can gain an idea of the scope and complexity of human cognition. The mind's fundamental intelligibility is suggested by successes of AI: its enormous richness and power are emphasised by AI's many current failures.

References

1. Lighthill, J. (1972). *Artificial intelligence: Report to SRC.* H.M.S.O.

2. Office of Technology Assessment, US Congress (1987). Artificial intelligence: a background paper. US Government Printing Office, Washington, DC. Cited in *The encyclopedia of AI*, (eds. S.C. Shapiro and D.Eckroth), pp. 1050–1 (1987). Wiley, New York.

3. Dreyfus, H.L. (1979). *What computers can't do: the limits of artificial intelligence.* Harper and Row, New York.

4. Michie, D. and Johnston, R. (1984). *The creative computer: machine intelligence and human knowledge.* Penguin, Harmondsworth.

5. Defense Advance Projects Agency (1983). *Strategic computing, new-generation computing technology: a strategic plan for its development and application to critical problems in defense.*, pp. 3–5. DARPA, Washington.

6. Sloman, A. (1984). Towards a computational theory of mind. In *Artificial intelligence: human effects*, (eds. A. Narayan and M. Yazdani). Ellis Horwood, Chichester.

7. Harmon, P. and King, D. (1985). *Expert systems: artificial intelligence and business.* John Wiley and Sons Inc., New York.

8. Commission for Racial Equality (1988). *Report of a formal investigation into St George's hospital medical school.* Commission for Racial Equality, Elliot House, 10/12 Allington St., London SU1E 5EH.

9. Powis, D. (1977). The signs of crime: a field manual for police. McGraw Hill, London. Quoted in Roshier, B. and Teff, H. (1980). *Law and society in England*, pp. 89–90. Tavistock Publications, London.

10. Dreyfus, H.L. and Dreyfus, S.E. (1986). Why skills cannot be represented by rules. In *Advances in cognitive science 1*, (ed. N.E. Sharkey). Ellis Horwood, Chichester.

4 Influencing the future uses of KBSs

KBSs are a growth industry. Even while we have been writing this report, more of them have been started, developed, or completed. Having described what they are, what they can and cannot do, and those of their benefits and dangers that we can foresee for ourselves, we are left with a final question: can we safely leave this growth industry to go its own way, or is there anything we, as a society, ought to be doing about it in order to maximise its benefits, or minimise its dangers?

The short-term future of KBSs is fairly well mapped out, by present-day research programmes such as Alvey and the policies of the few companies that are developing and marketing expert systems. But predictions beyond the next five years are very difficult. The field could develop in a number of directions, some beneficial, some harmful, and these directions will, in part, be determined by deliberate political and economic decisions.

4.1 Benefits

Clearly, the benefits of KBSs should be encouraged, just like the benefits of any other emergent technology that holds out comparable promises. But how that is to be done is another question. One possibility is to direct funds, both from Government and industry, towards specific initiatives with socially beneficial aims.

Doug Schuler of Computer Professionals for Social Responsibility has proposed a Responsible Computing Initiative (by analogy with the Strategic Defense Initiative). He describes a number of 'viewpoints':

Communication, Language and Literacy is designed for individuals and small groups. KBSs could be used to assist communication between 'communication-deprived' groups, such as illiterate people and those with language difficulties.

Resource Management is designed for larger groups and organisations. KBSs could be used to model scenarios that are important to the user, such as water distribution or crop allocation, or industrial uses such as factory scheduling. They could also help the groups manage their enterprises more thoughtfully through increased awareness of resources and the use of 'what-if' exercises.

Arbitration and Conflict Resolution is designed for nations and transnational organisations. The role of KBSs would be to assist negotiation, by supplying bookkeeping and other appropriate functions. Legal expert systems and arbitration models could facilitate peace through conflict resolution.

Richard Ennals, a UK researcher, in his book *Star Wars: a question of initiative*[1] has proposed a 'Strategic Health Initiative' funded by Government departments such as the DHSS and the Manpower Services Commission. It would build on research already being carried out, such as the Alvey DHSS Demonstrator project described in Chapter 1, and a large project KBS for molecular biology led by the Imperial Cancer Research Fund, and would fund projects concerned with preventative measures (such as dietary advice KBSs for the home and the workplace), scanning and detection of disease, coordination of resources, and aids for the disabled.

We would note that Government initiatives have been successful in the past in bringing a nascent technology to the stage where it can be taken up by industry and exploited in world markets. An initiative in health care would be popular and, in the long term, cost-effective.

> If such a programme were successful, the strategic results for the country would be spectacular. We could expect an improvement in the health of the population, with a cost-effective change of emphasis to prevention rather than cure, and a fall in the number of working days lost each year through illness. Industry could benefit from export sales of the resulting systems and the applications that followed in other sectors. The research community could benefit from the motivation of a continuation of work in 'advanced technology with a human face'.[1]

Some might argue that KBSs are not the only new technology that is worth supporting by subvention, that all such technologies should compete for funds on their relative merits, and that the best mechanism for allocating research funds is the traditional academic 'peer review'. Others may say that 'market forces' should dictate the scale and direction of KBS research. We have no special expertise which would enable us to advise about preferences between these mechanisms — or what might be the best mix of all of them — save to point out that markets work best when they are well informed, and so far at least very few people have more than the vaguest idea of the utility of KBSs. We therefore think that information about this technology should be made more generally available, and indeed that is one of the objectives of this report.

Allied to a general need for greater awareness is a specific one for education in the skills needed for designing and using KBSs. The 'stuff' of KBSs is knowledge, in a codified form, but nevertheless directly related to the knowledge of human experts, so there is an urgent need for people who are capable of acquiring knowledge from human experts, and of representing it in a formal notation. The skills are not those normally demanded of a computer scientist: interviewing techniques, an understanding of the psychology of human behaviour, an ability to use formal knowledge acquisition techniques such as protocol analysis and repertory grids, an empathy with the busy experts who may be unable to articulate their ingrained knowledge, and an awareness of the needs of those who will eventually use the KBS. A recent report on the commercial applications of KBSs from accountants Coopers and Lybrand [2] emphasises that they have 'misjudged the complexity of building "smart" AI-based system that incorporate appropriate expert knowledge':

> The field, when implementing successful, significant systems, draws extensively on a variety of disciplines such as philosophy, cognitive psychology, anthropology, semantics and linguistics, business and management sciences, in addition to systems science and cybernetics ... Most universities have faculty members in many of these disciplines with interests in various facets of the field, but few institutions have undertaken the creation of a new, integrated program to educate the professionals needed. The demand for individuals with such a background would be significant since they constitute the major bottleneck for extensive implementation of KBSs.

It may not be possible simply to graft such educational programmes on to the traditional degree structure of Universities and Polytechnics, since they require a commitment across Departments, with faculty from arts and sciences speaking the common language of cognitive science. Institutions that already have cross-disciplinary teaching have responded by setting up undergraduate study programmes in Cognitive Sciences, and MSc courses in Knowledge-Based Systems and in Cognition, Computing and Psychology. We believe that there is an urgent need for more such programmes of study.

4.2   Dangers

As well as promoting the benefits of KBSs, there is the problem of how
to minimise the potential dangers. This is in some ways more difficult.
Before one tries to tackle it, one must try to define the foreseeable dangers. The widespread introduction of KBSs may lead to social problems
— unemployment caused by automation, loss of human skills, loss of human responsibility for decision making — and we shall cover these later in
the chapter. Of the more direct dangers, the most important one is of a
KBS giving a 'wrong' answer, in circumstances where some real harm is
caused as a result. An example at the catastrophic end of the scale would
be the military KBS which wrongly believes that it has detected incoming missiles, and so unleashes a nuclear Armageddon. A less dangerous,
but nevertheless disturbing, example would be a pensioner who follows the
advice of a KBS, running on his home computer, to invest all his savings
in commodity futures, and rapidly loses the lot. Somewhere in between, a
KBS might suggest 20 aspirins 3 times a day as a suitable therapy for a
cold.

If such things happen, it will be for one or more of the following reasons:

- The knowledge base itself is 'wrong', such that data are either incor-
  rect or missing, e.g. it includes commodity futures in the class of safe
  investments, or it says 20 aspirins when it should say 2.

- The knowledge base itself is 'right', but the inferences drawn from
  it are wrong because of unreliable rules (ones which are not logical
  inferences).

- The knowledge base is right, but there is a mismatch between the
  meaning of a term as it it used by the system, and the meaning
  attached to that term by the user.

- The knowledge base is right, but its reasoning breaks down when
  presented with some unusual input contingencies which have not occurred to the designer.

- The decision criteria (or values) implicitly or explicitly built into the
  system may not be universally accepted. Thus it might be that giving
  the drug recommended by the program would prolong a patient's life,
  but at the cost of making that life unbearable.

Obviously, in a perfect world people who design, develop, distribute, and
sell KBSs should avoid such failures. In the real world, most of them will
indeed try to do that, but not all of them will always succeed — especially

as many KBSs are very complex entities which can never be fully tested before they are used. It is therefore a realistic assumption that there will always be some imperfect KBSs around, and that sooner or later a user of one of them will be at risk of suffering harm. One of the most important things for the world at large to realise about KBSs is that there is simply no way in which such risks can be entirely eliminated: all one can do is to try to reduce them.

4.3   Risks and remedies

In the field of KBSs, the degree of risk will generally depend on one critical question, namely the degree of expertise of the user in the expert system's own domain. A medical practitioner presented with advice by a KBS that the patient should swallow 20 aspirins 3 times a day will consign the disk and the manual to the waste paper basket, and address simultaneous rude letters to the distributor and the British Medical Journal. A lay person might not know better, take the aspirins, and leave his surviving dependants to find someone from whom to claim compensation. In this respect at least, the position will be no worse than if the family breadwinner had been killed through a latent defect in a motor car, since the law already provides a variety of remedies for situations of that kind — albeit only after the damage has been done.

In our example, the surviving relatives of the unfortunate victim of aspirin poisoning would need to find the High Street shop from which he had bought his KBS, and sue them for damages on the basis that the KBS was not fit for the purpose for which it was to be used, nor of merchantable quality — just as they would in the case of a motor car whose brakes had failed. The shop would doubtless bring in its supplier as a third party, and other parties would be brought in up the chain of distribution until one arrived at the original manufacturer. Claims for negligence might be added at some stage. Some or all of these parties might carry product liability insurance, and their insurers would therefore finance the litigation for them — particularly if it looked like becoming a test case. After a full and lengthy hearing which would doubtless include the evidence of many expert witnesses on all sides, the judge would have to decide where the blame for the calamity lay, and how much money, if any, should be awarded to the victim's estate. Even then, unless there were a legal obligation on all purveyors of KBSs to insure themselves against such awards — as there is for employers, and for the drivers of motor vehicles — the victim might not be able to recover the full amount awarded.

All that, of course, is a long, slow, uncertain and expensive process, and it can be started only after someone has been damaged. Some might say that this is good enough: life is full of risks and the State and its laws cannot be expected to protect everybody from everything; people should be encouraged to be self-reliant, rather than rely blindly on the help and advice of others (including KBSs), and it is therefore enough for the law to try to protect them from clear breaches of contract, and clear acts of negligence, but not from more than this.

4.4   Preventive regulation

Yet that is not the philosophy which has been followed in other fields. In nuclear power, for instance — admittedly a technology with a far greater potential for danger than any imaginable KBS except some of those in the military field, and in nuclear power itself — there is the most elaborate system of control over design, inspection and licensing, in order to ensure, so far as possible, before the event that 'nothing can go wrong'. It is not simply left to the victims of radiation damage to sue the operators after they have developed leukaemia. The same is also true for motor cars: here, we have comprehensive Construction and Use Regulations laying down safety standards with a high degree of detail, so that anyone who drives a motor vehicle which does not comply with them — let alone anyone who makes or distributes such vehicles — is automatically guilty of a criminal offence, even if no damage has been caused to anyone. In many such areas, the law seeks to prevent harm before it has been done, and does not just try to pick up the pieces afterwards.

Is there then a case for regulating the 'construction and use' of KBSs in some similar fashion? In principle, we believe that there is, but there are some obvious difficulties. First, KBSs are more varied than motor vehicles, and there are not many detailed safety regulations which could be applied to all of them. One cannot simply require that they must all have two independent braking systems, warning horns emitting a minimum of so many decibels, or two red lights visible only from the rear. Besides, the technology is still very young, and subject to many unforeseeable changes. It would therefore be far too early to begin to draft detailed regulations. But it is by no means too early to begin to think about the kinds of thing they might contain — and that is another purpose of this report.

Clearly, a distinction must be drawn between the one-off KBS developed by a large organisation for internal use only; KBSs intended to be used only by experts in their own domain; and KBSs available to the public at large, including potential users who have no prior knowledge of the domain. If there is to be regulation of any kind, it must apply with the greatest force to the third of these. Among the things which we think it ought to ensure are these (following Bundy and Clutterbuck):[3]

1. A clear statement, for each KBS, of who made it, on whose 'expert' knowledge it is based, and who will take responsibility if it causes any damage.

2. A clear statement, for each KBS, of what it claims to be able to do, within what boundaries, and subject to what limitations.

3. Clear statements, for each KBS, of the foreseeable risks and dangers it presents, as well as the risks and dangers inherent in the domain itself ('shares may go down as well as up').

4. Clear warnings about the dangers of relying on an impersonal system in a domain in which human experts are available ('if you don't feel well, you should go and see your doctor. All this system can do is to help you prepare yourself for that consultation. Do NOT use it to treat yourself: neither you nor this program knows enough about medicine.')

These seem to us to be the minimum statements, warnings, and disclaimers which should accompany any KBS sold on the open market, rather than to closed user groups of experts. These statements should appear not only on the label, or somewhere in the accompanying manual: they should be prominently displayed on the screen whenever the program is run, in a way which cannot easily be overridden.

Beyond that, we think there is room, and need, for generally accepted industry standards in various areas, especially in the 'interface' between the program and the user. Whenever the user is asked to supply some data, the request should be in clear, simple, and unambiguous language, leaving the minimum room for misunderstandings. Likewise, all answers should be suitably qualified ('on the information available so far, ... — but the answer might be different if ...') and if the system has concluded a number of results of similar confidence, then it should present them all, with the associated confidence, to the user.

4.5 Good practice

Clearly, it will be quite a few years before any such regulations are likely to be introduced. Parliament and Government have more urgent things on their agendas, and there is at present no perceptible public pressure for the regulation of KBSs. Meanwhile, however, there are several things that their present designers could do. First, they could devote far more effort to educating the public about this new technology. Secondly, they could proceed with the development of Codes of Practice among themselves. Until there are some laws about all this, these would of course be formally unenforceable. But they could have useful persuasive and educational effect meanwhile, and might later form the basis for legal regulations.

Whitby[4] has proposed a Code of Conduct for AI practitioners, drawn from the Code of Practice of the British Computer Society[5] and from the discussions of a working party of the Society for the Study of Artificial Intelligence and Simulation of Behaviour (SSAISB):

1. All professional persons working in AI should take all possible steps to ensure that customers, other professions and the public are not misled as to the degree of intelligence or competence possessed by AI systems. Descriptions and labels suggesting human attributes should be avoided where there is no technical justification for their use (as there might be if the system was intended primarily as a research tool in cognitive psychology, for example). Computer-based labels such as 'data', 'computation', 'processing' and so on should be used in preference to human-based labels, such as 'expert', 'inference', 'knowledge', and 'intelligence'.

2. Anybody required to work on an AI system in any capacity should have the right to know the ultimate purpose for which that system is intended.

3. Any professional person should be free to refuse to work on any AI system which they consider to be dangerous, likely to be misused or about which misleading claims have been or are about to be made.

4. Any professional person working on an AI system should be aware of the limits of his or her own competence and should not make misleading claims about his or her level of competence nor hesitate to obtain additional expertise from within AI or from other professions where appropriate.

5. Professionals working in AI should take all appropriate steps to update and maintain their level of expertise.

6. Full attention should be given to the sensitivity and security of data and the requirements of privacy and security.

7. A professional working on an AI system in any capacity should ensure that the interests of the end user are observed. In particular, AI

systems should not be introduced which will reduce or restrict the creativity, or the moral choices, of the end user.

8. Where an AI system is introduced into any human system, it should be the responsibility of the AI professional to ensure that a human or group of humans within the system should take moral and/or legal responsibility for the human consequences of any malfunction of the AI system.

9. An AI professional should introduce AI systems into human systems only where he or she is satisfied that both systems are organised in such a way as to ensure adequate management of the AI system.

10. An AI professional should ensure that he or she understands and can comply with the preceding nine requirements (or any others deemed appropriate), and should be prepared, if required, to give a legally binding statement that, to the best of his or her abilities, he or she has done so.

Not all firms or organisations could meet even the milder items above, but a voluntary Code of Conduct would nevertheless be useful — just as a statement by a firm that it has an Equal Opportunities policy is useful — in attracting recruits and providing a set of guidelines for research and development that is not merely based on market forces, or the general law of the land.

4.6 Automation at work

'Expert systems' sometimes called 'Knowledge Based systems' are capable of working like experienced and skilled staff.

Advertisements like this (from an AI company brochure) only foster a fear that KBSs may lead to unemployment and the downgrading of human skills. The practice of 'scientific management' or 'Taylorism' has led to the production line where manual work is accomplished by machines, with human workers as their adjuncts. Any exercise of human creativity, skill and judgement is rejected, as is the role of work in educating and developing the human spirit.[6] The labourer performs rigidly defined tasks, paced by machines. The fear is that KBSs will likewise pace or replace mental work.

We cannot say that this fear is groundless; indeed, the current preoccupation with 'performance indicators' and the 'objective measurement' of professions such as teaching and health care may create a context in which KBS systems are introduced to schedule and assess the work of teachers, nurses, or clerical workers. But the effect of KBSs on work cannot be predicted simply by studying the technology itself. The computer is a protean system; it can augment human activity, or it can replace it. At present, almost all existing KBSs are designed to assist problem solving, to improve the quality of human decisions, not to take over a task. Some autonomous KBSs are being designed, mainly for tasks that require a fast response based on the analysis of large amounts of data, but these are in specialised areas such as financial dealing or weapons control.

We suggest that KBSs should, wherever possible, be deliberately designed to complement human workers rather than replace them. Human creativity and judgement should be respected, and the control and scheduling of the task should rest with the human user. Working time 'freed' by the use of KBSs should be used for the development of other skills, especially those satisfying human skills which computer programs cannot perform. In this way workers' jobs and self-esteem will be protected.

4.7   The role of human beings

The decisions of KBSs should *always* in principle be challenged, for they work by consulting an internal model of the world (initially provided to them by the human programmer) and models are not necessarily accurate. It follows that one can always ask whether a model reflects the truth adequately or not.

We believe that by far the best safeguard against the dangers of any computerised expert system is to keep its activities under the constant supervision of an expert in the same domain, who will treat the KBS with appropriate scepticism and ignore its advice if it seems to him or her to be wrong. However, we have not been able to think of any practical way of ensuring that this obvious piece of common sense will always be followed: it is difficult to see, for instance, how it could effectively be required by law.

Again, all we can suggest here is the development of a much higher level of public awareness, so that there would have to be a major public debate before 'the human being was taken out of the loop' in any KBS with a serious potential for harm.

References

1. Ennals, R. (1986). *Star Wars: a question of initiative.* John Wiley, Chichester.

2. Wiig, K.M. (in press). Commercial applications. In *New technology and employment*, (ed. T. Bertold). North-Holland, Amsterdam.

3. Bundy, A. and Clutterbuck, R. (1985). Raising the standard of AI products. In *Proceedings of the ninth IJCAI*, pp. 1289–1294. Los Angeles.

4. Whitby, B. (1988). *AI: a handbook of professionalism*. Ellis Horwood, Chichester.

5. British Computer Society (1981). British Computer Society Code of Conduct. *BCS Handbook No. 5*. British Computer Society Publications, London.

6. Council for Science and Society (1981). *New technology: society, employment and skill*. CSS report, Council for Science and Society, London.

5 Conclusions and recommendations

5.1 Conclusions

In concentrating on the limitations and dangers of KBSs we have inevitably misrepresented the field. Research in artificial intelligence and cognitive science is, in a most literal sense, the pursuit of knowledge; it is an attempt to discover and formalise the rules, assumptions and processes of thinking. KBSs are the means by which these formalisms can be pinned down and made to perform. To try to capture human thought processes on a computer is a great and exciting challenge and one which is only just beginning. It is less than fifty years since the first computers were designed, less than thirty since Artificial Intelligence became a recognised area of study, and less than twenty since the first program recognisable as a KBS was produced. In that short time KBSs have become a billion-dollar industry. It is not surprising that, along the way, exaggerations have been uttered and mistakes have been made. The purpose of this report is not to restrict progress, but to point out some of the social consequences of this new technology, and to recommend ways in which it might be used to the benefit of society.

We were pleasantly surprised to find that, after much discussion, we could find few dangerous or undesirable applications of KBSs. We do not expect that the introduction of KBSs will be a direct cause of widespread unemployment, nor do we see it posing any special problems for privacy and civil liberties beyond those already posed by existing computer systems. Indeed, it may be that the act of representing knowledge and inference as explicit rules will help to uncover assumptions and bias in the minds of those humans on whom the program is modelled.

In our view, by far the most worrying potential application is the 'autonomous decision making system', a KBS that not only makes decisions but also commands machinery to act on them, and this danger is at its greatest in the military field. Unfortunately, that field is — understandably but regrettably — surrounded by such a degree of secrecy that it is extremely difficult to mount any informed public debate in it. Our worries are compounded by the fact that it has so far been little discussed, and that the pressures against informed public discussion of it are so great that this state of affairs may well persist for a very long time — long enough for a major catastrophe to happen meanwhile.

5.2 Recommendations

1. The general level of public awareness about the applications and social implications of KBSs is low. Computer literacy is now an essential part of literacy, and should be taught as such. We therefore recommend that education about KBSs should be an integral part of literacy teaching in schools. This education can start in the primary school, with the construction of simple expert systems and demonstrations of the uses to which different types of knowledge can be put.

2. There is an urgent need for more undergraduate and graduate interdisciplinary study programmes in cognitive sciences and in KBSs.

3. We suggest that a Government initiative in the application of KBS technology to health education and health care would be popular and, in the long term, cost-effective.

4. A KBS should, wherever possible, complement human workers rather than replace them. It should be designed to explain its reasoning, allowing the human user to direct the task and exercise judgement in interpreting the results.

5. We think it is undesirable to substitute a computer for a human function, such as giving psychiatric help, that should involve respect, understanding, empathy or love between humans. (This does not preclude the use of such systems as an aid in the consultations between the human professional and client.)

6. In all discussions by governments concerning deterrence and armaments, great attention should be given to the immense hazard which is presented by the possible malfunctioning of military command and control programs and autonomous decision-making systems.

7. We propose that the Data Protection Act should be extended to entitle people to know not only what data are held about them for what purposes, and to whom they are disclosed, but also the rules by which they are processed.

8. There may, in the future, be a need for statutory regulation of KBS standards.

9. The vendors of KBS systems should be encouraged to adopt a Code of Practice similar to that proposed in Chapter 4.

10. Meanwhile, there should be a legal obligation on all purveyors of KBSs to insure themselves against claims for damages by users.

11. There is a need for industry standards in various areas, especially in the 'interface' between the program and the user.

12. Professional associations, such as the Society for the Study of Artificial Intelligence and the Simulation of Behaviour (SSAISB) and the Expert Systems group of the British Computer Society, should be encouraged to adopt a Code of Conduct for their members.

13. We do *not* recommend legislation to set up a formal body for KBS practitioners with restricted membership, although this might be appropriate in the future.

Bibliography

Bishop, P. (1986). *Fifth generation computers: concepts, implementations and uses*. Ellis Horwood, Chichester.

Boden, M.A. (1987). *Artificial intelligence and natural man*. 2nd ed., expanded. MIT Press, London.

Bundy, A. and Clutterbuck, R. (1985). Raising the standards of AI products. In *Proceedings of the ninth IJCAI*, pp. 1289–94. Los Angeles.

Daedalus (Joint American Academy of Arts & Sciences) (1985). *Weapons in space*. 2 volumes: *Concepts and technologies* (spring 1985) and *Implications for security* (summer 1985).

Dreyfus, H.L. (1979). *What computers can't do: the limits of artificial intelligence*. Harper and Row, New York.

Ennals, R. (1986). *Star Wars: a question of initiative*. John Wiley, Chichester.

Feigenbaum, E.A. and McCorduck, P. (1983). *The fifth generation: artificial intelligence and Japan's computer challenge to the world*. Addison-Wesley, New York.

Feigenbaum, E.A., McCorduck, P. and Nii., P. (1988). *The rise of the expert company*. Macmillan, London.

Gardner, A. v.d.L (1987). *An artificial intelligence approach to legal reasoning*, MIT Press, Cambridge, Mass.

Hand, D.J. (1985). *Artificial intelligence and psychiatry*. Cambridge University Press, Cambridge.

Handy, C. (1984). *The future of work: a guide to a changing society*. Basil Blackwell, Oxford.

Hayes-Roth, F., Lenat, D.B. and Waterman, D.A. (1983). *Building expert systems*. Addison-Wesley, Reading, Mass.

Jahoda, M. (1984). Unemployment: curse or liberation? In *New technology and the future of work and skills*, (ed. P. Marstrand), pp. 19–25. Frances Pinter, London.

McCorduck, P. (1979). *Machines who think*. Freeman, San Francisco.

Michie, D. (1982). The social aspects of artificial intelligence. In *Machine intelligence and related topics*, (ed. D. Michie), ch. 27. Gordon & Breach, New York.

Parnas, D.L. (1985). Software aspects of strategic defense systems. In *Scientific American*, Sept-Oct 1985, pp. 432–40.

Parnas, D.L. (December 1985). *Software and SDI: why communications systems are not like SDI*. Senate Testimony. Available from: CPSR, PO Box 717, Palo Alto, CA 94301, USA.

Rosenbrock, H.H. (1984). Designing automated systems: need skill be lost? In *New Technology and the Future of Work and Skills*, (ed. P. Marstrand), pp. 124–32. Frances Pinter, London.

Sharples, M. (1985). *Cognition, computers, and creative writing*. Ellis Horwood, Chichester.

Sloman, A. (1984). Beginners need powerful systems. In *New horizons in educational computing*, (ed. M. Yazdani), pp. 220–34. Ellis Horwood, Chichester.

Susskind, R. (1986). Expert systems in the law: a jurisprudential approach to artificial intelligence and legal reasoning. In *Modern Law Review*, 49, pp. 168–94.

Susskind, R. (1987). *Expert systems in law*. Clarendon Press, Oxford.

Waterman, D.A. (1985). *A guide to expert systems*. Addison-Wesley, New York.

Weizenbaum, J. (1976). *Computer power and human reason: from judgement to calculation*. Freeman, San Francisco.

Wenger, E. (1987). *Artificial intelligence and tutoring systems*. Morgan Kaufman, Los Altos, Ca.

Whitby, B. (1988). *Artificial intelligence: a handbook of professionalism*. Ellis Horwood, Chichester.